WHAT PEOPLE ARE SAYING

A Godsend! This book is destined to save millions from unnecessary misery from mold, offering an effective solution that is both simple and inexpensive. Use it. Share it with your friends and relatives. Tell everyone you meet. Tell those who have chronic ailments that remain unexplained or unsuccessfully treated by doctors that maybe their problem is mold, even if they don't think there is any mold in their home or workplace. Tell them to read this book and apply the Close Protocol. It could restore their health and rekindle, for them, a new joy in living. It could even save their life.
David Stewart, PhD, DNM, IASP President, CARE (Center for Aromatherapy Research and Education) Author of the books *Chemistry of Essential Oils Made Simple*, and the internationally acclaimed *Healing Oils of the Bible*

Informative, concise, I thoroughly enjoyed every word. The world needs this information on fungus eradication to help improve the health of millions. Dr. Close, with the insistence of his aromatherapist wife, has stumbled upon a miracle that can aid the world. I was reading [the book] while attending a convention for holistic nurses, so I of course told many people who came to our booth all about your research ... everyone who purchased oils at that convention did so as a result of telling them about your research on the Thieves® blend of essential oils used in your tests. I definitely want to carry your book in our classrooms. This book is going help people all over the world.
Linda L. Smith RN, MS, CCA President, Institute of Spiritual Healing & Aromatherapy (ISHA) Author of *Healing Oils Healing Hands*

The clearest, most comprehensive, fascinating read. Superb! The Close's educational book identifies an age old problem using personal

accounts along with scientific research. If you suffer with allergies, asthma, ear infections, colds, heart disease, or any malady -- this is a must read -- and also for the professional, the homeowner, the property manager and developer. A phenomenal breakthrough, offering humanity a powerful solution to remediate deadly mold with a non-toxic, natural, 100% effective essential oil treatment – so safe you can eat it. This book can literally save your life.
Sabina Mary DeVita EdD, DNM, RNCP Director of the Institute of Energy Wellness Studies (IEWS) Author of the books *Saving Face* and *Electromagnetic Pollution*

I highly recommend this book. I couldn't put it down. I have had the opportunity to see a lot of mold victims, and while there is much confusion in the literature about the effects of mold on people's health, we need to know how to find it and how to treat it. This book was very enlightening.
Norman L. Dykes, MD, APMC, Internal Medicine/Primary Care Wellness & Prevention, Lafayette, LA

Great book! Dr. Close has provided us with a meticulously documented and highly accessible study of how to use essential oils to banish mold. It is rare to find a scientist who is willing to be so thorough in documenting and testing a novel treatment approach. His work will provide a foundation for wider use and acceptance of essential oils in practical applications. This work also serves as a model for future scientific evaluation of emerging answers to our most pressing health crises.
Dr. Benjamin Perkus, Licensed Psychologist and Health Coach

The sheer amount of data is very conclusive. Proving that this essential-oil blend effectively eliminates fungal spores from the air and prevents them from returning for weeks or

months in living spaces is indeed an important breakthrough discovery. I hope that this information can be incorporated in standard operating procedures in this industry. It will prevent many health problems from which many of us have been suffering. This is indeed an apt area for research that affects us all in our everyday life. The anecdotal information on the health of inhabitants of the buildings was very useful. Perhaps additional research could be done that uses a more standardized before-and-after documentation of health problems in relation to specific fungi found in the spaces in order to help establish a definite cause-and-effect relationship.
Don Miles, PhD, Clinical Microbiologist

If you suspect you have a mold problem or mold-related illness, read this book. If you *know* you have a mold problem, don't hire a mold remediation contractor until he has read this book and agrees to follow Dr. Close's 10-step protocol. Following Dr. Close's recommendations saved me nearly $1500. The best news, though, is that my cat's breathing problems, which had been steadily and mysteriously worsening, were almost nonexistent the week after we eliminated all the mold. This book is going to help many, many people.
Connie Bennett, Homeowner

This book reveals a unique method of mold remediation, especially when considering the health of a building's occupants. The research provides compelling evidence that therapeutic grades of essential oils can kill molds with **NO** residual toxic outgasing effects and the efficacy of the products are very important, and offers the first products on the market that are also beneficial for the human body. No other Fungicide, Sporicide, Sanitizer, etc., can say that. This is a great solution for use without the fear of exposing bodies to harsh products such as bleach, ammonia, petroleum-based, and other compounds. Only these therapeutic grades of essential oils have been proven to work against molds. Other

synthetic essential oils will be tried/used, but they will not work!
Richard F. Progovitz
Certified and Registered Mold Inspector and Remediation Contractor
Author of *Black Mold - Your Health and Your Home*

I want you to know, I thoroughly enjoyed this book. It is so readable and especially good for the "unscientific" person. I particularly liked the fact that the information takes one from realizing mold is a factor in one's life, clear through completion of the remediation project in a clear and concise manner.
Elaine Pinar, LPN

I certainly enjoyed reading your book. Mold is quickly emerging as one of today's most highly feared health threats. The tag-team authors of Dr. Ed Close and Jacqui Close successfully move the story past the fear stage and present convincing documentation and compassionate stories about this revolutionary mold-busting technology. We can all rest a lot easier, if we are educated with the potentially life-saving information presented in this book.
Tim Walker, Professional Geologist (PG)
Northwest Florida Water Management District

This book is meticulously researched and demonstrates a simple, environmentally friendly, safe way to remove and eliminate toxic molds by using essential oils. Everyone who wants to restore and maintain optimal health will be served well by reading this book."
Edgar L. Miesner, DC

Anyone seeking irrefutable, scientific evidence of the powerful effectiveness of treating mold with therapeutic-grade essential oils has it in their hands with these well-documented case studies. Dr. Ed Close and Jacquelyn Close, RA, have performed an invaluable service by a) doing this intensive research and creating protocols and b) meticulously recording the results in this book for us to use to teach others how they can remediate the serious problem of molds–a root cause of many debilitating illnesses and diseases–without resorting to chemicals, which exacerbate the problem. Well done!

Some facts that I found particularly interesting:
* that in nearly every case, people's symptoms greatly improved <u>immediately</u> following the essential oils treatment of their surroundings that had been mold-infested.
* that there were higher mold counts in a building that had no <u>visible</u> mold than in those where it was obvious
* that remediation did not work as well as when the Thieves Household Cleaner was diluted with water.
* that Case #16 created a great testimonial–comparing the results of the woman's treatments with allopathic medicine and the essential oils and oil products.

Joy Linsley, Gold Young Living Distributor

The Mold is Gone, and My Health, Life and Happiness Have Been Restored

My husband and I live on a farm in a beautiful historic area just west of the Mississippi River in Perry County, Missouri. We've lived here for 26 years and raised our family here. Several years ago, I started having strange health problems, including constant aching-all-over, flu-like symptoms, fatigue, miscarriages, and alarming mental decline. With an education in health care and nursing, I worked as a nurse until my illness and memory loss made it impossible for me to continue. I was diagnosed with "seizure activity." My doctor prescribed a number of drugs over the years, but nothing seemed to work. The cause of my debilitation was a mystery.

Several months ago we noticed black mold in our basement. After coming in contact with it, I became very ill and had to be rushed to the hospital with a strange bout of allergic reactions. God answered our prayers and I arrived at the emergency room just before my airways closed off entirely. The Doctor thought it was an adverse reaction to medications, but I felt that it was due to my contact with the mold in our basement.

A friend at church heard about my story and told me about the research Dr. Edward Close was doing using non-toxic essential oils to deal with toxic mold. I was so excited when she told me he would be speaking in Perryville.

After his talk, I asked Dr. Close if he could help us. He came to our home, did an inspection and took samples which showed we had *Satchybotrys*, along with several other species of mold growing in our home and basement. We followed Dr. Close's protocol using Young Living's Thieves® oil and Thieves® Household Cleaner, and my symptoms began to improve almost immediately!

Thanks to Young Living Essential Oils and Dr. Close's protocol, I have my life back and my health has been restored.
Thank you so much,
Debbie and Steve Thieret, Homeowners

What a wake-up call! As a chiropractic physician, I can relate that many of the chronic respiratory symptoms and health problems of my patients are related to exposure to mold spores. This book brings it all together and identifies a non-toxic treatment that is highly effective.
Gary J. Bridges, DC

We are both teachers and diffuse therapeutic grade essential oils in our classrooms. We have seen definite improvement in our students' attendance and seemingly better health. Last year a flu was going around, and in the classroom next to me, the teacher and all but five students were out sick. I had used the Thieves Spray® on the desks and doorknobs, in addition to diffusing essential oils in my classroom, and all twenty-three of my students were present with me. We also know several people in the case studies cited in this book, and know, personally, how these therapeutic grade essential oils have changed their lives for the better.

In this book is scientific proof that Therapeutic Grade Essential Oils can reduce, if not eliminate toxic levels of mold. Now, with the simple protocol of cleaning and diffusing, 90 to 100% of all toxin-producing and infection-causing species of mold can be eliminated. It Works!"
Dr. Harry W. Pickup III, DMA, and Linda Reinwald-Pickup, MS Edu

Ed and Jacqui Close's latest book, *Nature's Mold Rx*, is the definitive scientific work on the remediation and mitigation of toxic mold found in homes, institutions, vehicles, and business environments. Using Thieves® essential oil blends from Young Living Essential Oils in the ways described in the book not only kills toxic mold, but safely prevents its recurrence in a variety of environments. It has been known for several years that the overgrowth of mold and other fungi in the lungs and sinus cavities causes disease by creating a heavy load on the immune system that excessively raises serum globulin levels, thus lowering serum albumin levels (and thus immunity) in both humans and animals (Sources: Dr. Kenneth Seaton, the Mayo Clinic studies). By taking the toxicity load off the immune system, many allergies, diseases, and immune dysfunctions simply go away. The Close book is a must read for all health and healing practitioners, especially allergists, professional toxic- mold remediation firms, and, of course, the general public.
Bob Krone, Certified Lymphologist

My daughter, Jessica Grady, purchased a business where we discovered mold growing in some carpeting and drywall. I suspected that this mold was causing the health problems that she was experiencing, but she was blaming her constant fatigue and congestion on the stress of the business demands and her new pregnancy. When she went on vacation for a week she felt so much better, and then as soon as she returned to her business, all of her symptoms immediately returned. This sent up the red flag to me that this mold had to be tested right away, especially due to my fears about the effects of toxic mold on her unborn baby. I got on the internet to research the solution for this potentially serious problem, but was very concerned because information I found said that toxic mold has been shown to be very difficult if not impossible to rectify. So I prayed for guidance.

Then, I discovered a posting by Ed and Jacqui Close on an email chat-group relating their successful experiences utilizing a safe, 100% pure, natural blend of essential oils to destroy toxic molds, including *Stachybotrys*. I contacted them immediately and requested Ed take samples for testing.

The bad news: We not only had *Stachybotrys* present in large amounts, but also two other strains of toxic *Aspergillius*.

We followed Dr. Ed Close's instructions on diffusing the Thieves® essential oil blend in the area and spraying the Thieves® Household cleaner, directly on the mold. And we arranged for the contaminated carpeting and drywall to be professionally removed by a mold-remediation specialist right away. But, even before the contractor was able to accomplish this, in fact, within the first day of using the Thieves® oil treatment, my daughter experienced a big improvement in her symptoms. Within a short time, all of her symptoms disappeared and she was so grateful to be feeling well again. Both of us are extremely grateful that God directed us to Ed and Jacqui Close for this miraculous solution.-- With Gratitude,
Evelyn and Jessica Grady

Nature's Mold Rx

A Breakthrough Discovery

Edward R Close, PhD,
Professional Engineer
and
Jacquelyn A Close,
Registered Aromatherapist

Foreword by David Stewart, PhD, DNM, IASP

EJC Publications
First Edition, 2007

the Non-Toxic Solution to Toxic Mold

EJC Publications
PO Box 368
Jackson, MO 63755
Phone: 1-877-756-6753
email: info@ejcpublications.com
www.ejcpublications.com

Cover Art: *Aspergillus versicolor* fruiting structure photo by Dennis Kunkel Microscopy, Inc., www.denniskunkel.com
Cover design by: Silver Moon Designs, www.affordablewebpages.com

---

LCCN 2007933215
ISBN 978-0-9785616-1-1
Close, Edward R and Jacquelyn A
      Nature's mold rx: The non-toxic solution to toxic mold; a breakthrough
         discovery/ by Edward R. Close and Jacquelyn A. Close; with a Foreword by
         David Stewart, PhD. – 1st ed.
      Includes bibliographical references and index.

1. Aromatherapy. 2. Molds (Fungi) – Control. 3. Allergy – Popular works.
4. Essences and essential oils – Therapeutic use – Testing. 5. Molds (Fungi) – Health aspects. 6. Indoor Air Pollution – Popular works. 7. Dwellings – Maintenance and repair.

Printed and bound in the United States of America

DEDICATION

This book is dedicated to all who suffer from chronic allergies, chronic colds and other health challenges associated with mold exposure; to all who seek answers, knowledge and understanding; and to all who stand in awe of the provisions made for us by our loving Creator.

ACKNOWLEDGMENTS

We wish to acknowledge and extend our heartfelt thanks to the following people for their generosity of spirit and nurturing editorial comments and suggestions. The inspiration, assistance, training, encouragement, and support received from each is most sincerely appreciated.

Connie Bennett
Gary J Bridges, DC
Sabina DeVita, EdD, DNM
Lana Dietrich
Norman L Dykes, MD
Mary Lou Kowaleski, MS
John Kraemer, PhD
Joy Linsley
Edgar L Miesner, DC
Donald Miles, PhD
Benjamin Perkus, PhD
Dr. Harry W. Pickup, III, DMA
Linda Reinwald-Pickup, MS Edu.
Elaine Pinar, LPN
Richard F Progovitz, Mold Remediation Consultant
Cheryl Progovitz, RN
Linda L Smith, RN, MS, CCA
David Stewart, PhD, DNM, IASP
Tim Walker, PG
Claudia Zimmerman

CONTENTS

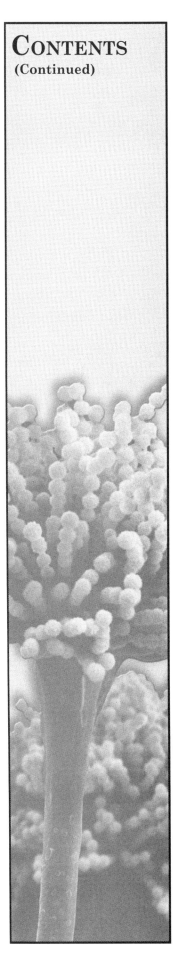

CONTENTS
(Continued)

CONTENTS
(Continued)

PART FIVE — APPENDICES AND RESOURCES

FOREWORD

David M. Stewart, PhD, DNM, IASP

Dr. Ed Close and I have been best friends for most of our lives. We met as freshmen at Central Methodist College, in Fayette, Missouri, in 1955 and were college roommates. As long as I have known Ed, he has always been on the forefront of scientific, mathematical, and philosophical thinking – publishing books and articles, teaching classes, and forging new frontiers.

I have also known his wife, Jacqui, for several decades and have learned much from her in the field of aromatherapy, in which she is a registered and recognized authority. Jacqui, in her life and thinking, has always been far ahead of her time. It was the unique union of Jacqui's insight into applied aromatic science and Ed's experience as an innovative environmental engineer that gave birth to the breakthrough discovery that essential oils are the answer to the problem of toxic mold.

Toxic mold in the human environment has always been an unsolved problem. Until now.

What Dr. and Mrs. Close have developed is a 10-step protocol for using a proprietary blend of oils that kills toxic mold effectively, safely, and inexpensively. I call it the "Close Protocol." The essential oil blend they used consists of oils of clove, cinnamon, lemon, eucalyptus, and rosemary. It was the second of two blends they initially tested, and, for that reason, in this book they often refer to it as EOB2.

Through a series of well designed scientific experiments in a variety of buildings, Dr. Close has proven that this blend can kill 100% of the most toxic molds. It not only kills the mold but actually removes the mold from the air, including the dead mold. This is significant because dead mold is also toxic when inhaled.

Conventional methods for dealing with mold are costly, disruptive, and ineffective. They often involve major removal of walls, floors, and whole sections of a house or building and, in extreme cases, even abandonment of the property. The chemicals used are toxic to people, plants, and animals, which poses yet another hazard in addition to the mold. No conventional method is 100% effective, like the oils. Hence, even with 99% effectiveness, there are still some live mold spores remaining that can eventually multiply into billions, thus re-infesting the property. It is common for mold to start re-establishing itself within days of applying the usual chemicals currently used.

Applying essential oils to combat mold is not only harmless to people, it is actually beneficial. Instead of leaving a hazardous residue in your home, as the chemical approach does, the blend of essential oils used in these tests remains active in the environment for days and prevents future fungi from re-growing. In addition, the health-promoting properties of these oils can boost your immune system and reduce your susceptibility to bacteria and viruses for the long term.

Untold numbers of people suffer from flu-like symptoms, mental disorders, pneumonia, allergies, unexplained aches and pains, headaches, loss of hearing, skin disorders, and even more serious illnesses like cancer, that share a common cause: toxic mold. Because doctors are not trained in identifying environmental fungi as a source of sickness, these ills go misdiagnosed and mis-prescribed with drugs and procedures that never solve the problem and usually make it worse. In many cases, the lack of recognition on the part of physician and patient that mold is the culprit results in years of wasted medical expenditures, unnecessary suffering, permanent disabilities, and even death.

The source of the mold can be in houses, businesses, retail stores, warehouses, apartments, offices, workplaces, elder-care facilities, and even hospitals. Any inside place with a dampness or moisture problem can become a fertile home for fungi. Among the most valuable portions of the Closes' fact-filled book are the twenty case histories of serious mold infestations successfully solved with essential oils. These cases include every type of enclosure from antique homes, government offices, senior citizen facilities, and hospital operating rooms to bars, restaurants, and newborn intensive-care units.

Besides the Close DVD on the subject, this book is the only source for this information. The book contains a great deal more information than the DVD. This is pioneering science. These are the very first published studies on this topic. This is the first practical protocol ever devised for solving mold problems with essential oils.

In addition to the scientific solidarity of the Close Protocol for mold mitigation, the solution offered in this book is special in another way. Ed and Jacqui are also deeply spiritual people who have long lived a life of daily prayer. They both acknowledge the divine inspiration for their findings and the spiritual guidance they received that led

to their discovery. In their book, as well as in the DVD they have produced, they are passing on to the world a gift they feel was given to them from God.

This book is destined to save millions from unnecessary misery from mold, with an effective solution that is both simple and inexpensive. Use it. Share it with your friends and relatives. Tell everyone you meet. Tell those who have chronic ailments that remain unexplained or unsuccessfully treated by doctors that maybe their problem is mold, even if they don't think there is any mold in their home or workplace. Tell them to read this book and apply the Close Protocol. It could restore their health and rekindle, for them, a new joy in living. It could even save their life.

David Stewart

An internationally renowned teacher and retired United Methodist Pastor, Dr. David Stewart holds degrees in math, physics, earth sciences, natural medicine, and aromatic science. He is President of the Center for Aromatherapy Research and Education (CARE), which offers seminars and books on the science and practical application of essential oils. Formerly a professor at the University of North Carolina at Chapel Hill and at Southeast Missouri State University, he currently serves on the faculty of the Institute for Energy Wellness Studies in Brampton, Ontario, Canada. Author of sixteen books and more than three hundred articles, Dr. Stewart's latest book is entitled: The Chemistry of Essential Oils Made Simple. *His best-selling book,* Healing Oils of the Bible, *has garnered international acclaim.*

INTRODUCTION

IS THERE A NON-TOXIC SOLUTION TO TOXIC MOLD?

A young mother contacted us. Her twenty-month-old son has black mold coming out of his ears. How did she know this? She had taken her baby to an ear, nose, and throat (ENT) specialist, who had done a culture. The black things coming out of her son's ears were colonies of *Aspergillus niger* spores, one of several species of fungi that are commonly known as toxic black mold. She also said she had the same toxic mold growing in her home.

A former nurse called us. In her forties and on disability, she had lost her health, her job, and most of her hope of ever having a normal life again, but in desperation she reached out one more time and asked us, "Can you find out if mold is growing in my home?" Within a few days of treating the mold with the protocol we developed (presented in Part 4 of this book), she called and said she already felt better. A few weeks later, she called to tell us she finally had her life back again.

A woman in her sixties called. She has been diagnosed with chronic fatigue syndrome, fibromyalgia, and was chronically depressed. She had lost the hearing in one ear. She had what looked like long, deep cuts on the tips of her fingers. She was taking many prescription medications (the list filled two pages) and had not been able to travel by car more than two miles from her home for years without

suffering from total exhaustion. She has heard about our work with mold and wonders if toxic mold might be contributing to her problems. Testing showed she had several mold species growing in her home. Within a month of treating the mold with our non-toxic, eco-friendly essential oils and offering some additional suggestions to support her body, her fingers had healed, she was able to travel seventy miles (one way) by herself, visit with a dear friend she had not seen for years, and return home full of energy. Her doctor reduced many of her medications and took her off several of them. Two months later she called to say her therapist, whom she had been seeing every week, said she was doing so well that she need not come back for three months. She was so excited with the turnaround in her life, and said getting rid of the mold and supporting her body's natural healing capabilities were the answer to her many years of prayers for help.

People ask us how we got started doing work with mold, and the truth is it came to us. Owning a small environmental engineering firm in Southeast Missouri meant we were the company people called for many different types of environmental problems.

Ed has been in the environmental industry since the 1960s and Jacqui since the 1970s. Ed had worked for the US Geological Survey

2

(USGS), and he and Jacqui met when a consulting firm she worked for sent her to the USGS to pick up maps. The company she worked for had been founded by two former USGS employees. Marriage, a child, and Ed's move into working for large environmental consulting firms took us around the world and back. In 1992, we returned to Ed's home state of Missouri to be near his aging mother. When the firm Ed was working for told him he would have to move to Chicago or Kansas City, Ed and Jacqui decided it was time to stop moving and open their own firm. So, in 1995, with two clients, they opened Close Environmental Consultants. Ed already had thirty years experience in the environmental industry, a PhD in environmental engineering, and had been an advisor to more than ten Fortune 500 companies. Jacqui had grown up in family-owned businesses. Her business and accounting background proved most helpful in their engineering business.

Ten years later…
It was a cool, late October day in 2005, when a long-standing client of Close Environmental Consultants contacted us and asked that we do third-party sampling for mold in an apartment complex their company had purchased and was renovating. The apartments had been flooded, then evacuated, and later put up for sale. Closed and vacant for some time, mold had run rampant in the buildings. Our client was renovating the apartments, and his lender required that the apartments pass a mold inspection.

Mold and allergies are a big problem in this part of the world, but most people just accept it as a part of life along the Mississippi River. When Ed went to the apartments to do the sampling, most of the water-damaged, mold-infested material had already been removed, and a cleaning company had already been through the apartments and cleaned everything with a solution that they told the new owners would kill any mold. The sampling results showed that numerous molds

and especially molds of concern, which are commonly known as toxic molds, were still extraordinarily high when compared to mold levels found in the outside air and also to levels typically found in the surrounding region.

Our client referred Ed to the cleaning company to find out what had been used during the cleaning activities, who said they had used a solution their supplier told them would kill mold and anything else that was harmful to human health. They said: "It's the same product used for hospital clean rooms."

Based on the results of sampling, the product had not killed the mold. Or, if it had, the mold had already re-established itself.

It is well known to people working in the mold-remediation industry that bleach, as well as most cleaners and industrial-strength fungicides commonly used in mold remediation, are not effective in reducing or eliminating mold spores for more than a few days. Additionally, most of the products commonly used are more toxic, more hazardous, and more dangerous to human health than the mold spores and their associated toxins. After reviewing the documentation and the Material Safety Data Sheets (MSDS) for the products the cleaning company had used, Ed learned they were bactericides, and suggested they acquire an industrial-strength fungicide and re-treat the buildings. They did. A second round of tests showed that the levels of mold were still extraordinarily high (See Case Study No. 1).

Ed was discussing this situation with Jacqui (who had discovered essential oils in 1995 and became a registered aromatherapist in 2001), and she asked,

"What if there was a non-toxic solution to toxic mold?"

2*Introduction*

Ed said there were a few options that offered that, but they each had significant drawbacks. Jacqui urged Ed to talk to the client about allowing him to do a test in several of the apartments to see if essential oils would work. Essential oils are known to be antifungal, and the ones that are commercially available are non-toxic at the levels used in the Protocol (See Part 4).

At first, Ed resisted this suggestion. That was Jacqui's other business. It had nothing to do with engineering, and he was a professional engineer. To be honest, Ed did not feel comfortable suggesting to his clients that they use essential oils to treat a toxic mold-infested building. More than that, Ed doubted it would work and did not want to suggest anything without sufficient scientific basis to do so.

Jacqui was not deterred by the doubts and reservations. She knew about the power of therapeutic grade essential oils. She had worked with essential oils for ten years and had discovered that while essential oils were highly concentrated, not all essential oils were of sufficient quality to be reliable. Indeed, few companies in the United States sell high quality, therapeutic grade essential oils.

Jacqui brought Ed information on scientific studies showing the antifungal properties of essential oils and showed him anecdotal evidence of how people used the oils against mold. Ed reviewed the materials and asked lots of questions. Jacqui showed Ed a notation on the bottle's label that said the oils were approved by the United States Food and Drug Administration (FDA) as a food supplement. To Ed, that meant they were safe, and he had personally experienced positive health benefits from the essential oils proposed for the tests.

Lacking any other viable and economically feasible alternative, Ed decided to approach his clients. He was careful to be upfront with them and told them he had no idea if this

option would work; this would simply be a test. Ed assured them that he would do everything he could to set up the test so that the results would show whether this was a viable alternative or not. Having been a research scientist early in his environmental career, he had a good background and preparation for accomplishing this task. The clients were willing to give it a try. The rest of this story and the results of the tests can be found in Part 3: The Case Studies.

That was the beginning, the impetus behind the research that is included in this book.

So, we began the research with the question for which we truly did not know the answer:

Can essential oils be used to effectively remediate mold in buildings?

The data presented in this book depict the results of that first test case and nineteen additional case studies (see Part 3). Analyses for all samples collected were performed by Environmental Science Corp., an independent, EPA certified, environmental testing laboratory located near Nashville, Tennessee. Samples were collected by Edward R. Close, PhD, Professional Engineer (PE) in the State of Missouri and Principal with the firm Close Environmental Consultants. A proprietary blend of organic, therapeutic grade essential oils and a household cleaner containing the same proprietary blend of essential oils were used in the tests reported in this book. The essential oils blend and the household cleaner are produced by a company in Lehi, Utah. Field tests were designed by Ed Close to determine whether the proprietary blend of essential oils was effective in remediating mold in buildings. Results of all case studies are presented in Part 3 of this book.

Recently, we were contacted to do mold sampling in a home where a woman in her sixties and her ninety-year-old mother had lived alone for many years. This house had been their family home since 1926. All other close relatives had either died or moved out of state.

> Research continues to indicate links between mold and every sort of disease, including the major killers: cancer and heart disease. Yet most people do not consider mold a serious threat.

When the mother's health failed, conditions in the home deteriorated because the daughter had to spend almost all of her time caring for her mother, cooking for her, feeding and bathing her, taking her to the doctor, and making sure she took her medications on time. There was precious little time for cleaning.

The situation became more and more stressful. The daughter's health began to suffer. She had memory lapses and found it hard to express herself. She was often confused. When she went to her family doctor, she was told it was probably just aging and stress. Her doctor prescribed vitamins and antidepressant drugs, which she said did not help but she kept taking them because her doctor had prescribed them.

The mother passed away. Relatives came for the funeral, and one sister decided to stay longer to help clean and tidy up the house. After spending a night in the house, the sister began to experience headaches. After the second night, the sister woke up with a red and swollen eye. Noticing a lot of black mold growing on the windows in the bedrooms, she decided to do some research into the health effects of mold, and soon afterward she contacted Ed to do sampling. When Ed met with the woman who had lived in the house for years, she could not complete a sentence. She could barely put three words together

before losing her train of thought. The rest of this story is presented in Case Study No. 20, found in Part 3.

It has been known for more than a century that airborne bacteria and viruses cause colds and influenzas. Research is now showing that inhalation of airborne mycotoxins (toxins from mold) and mold spores can cause respiratory infections, asthma, and all the symptoms of colds and flu, and links have even been found between mold and every sort of disease including the major killers: cancer and heart disease. Yet most people do not consider mold a serious threat. And most doctors do not test for mold spores or associated mycotoxins in the blood, and do not even ask their patients if they have mold growing in their homes.

Across the country and around the world, people are falling ill, suffering, and even dying from illnesses for which their doctors cannot, do not, or rarely identify a cause.

Fungi are ubiquitous in the environment. Many of them are known to produce toxins that cause health problems ranging from minor irritations to life-threatening conditions. While mold may not always be the cause, it may contribute to or exacerbate illness. Yet, because molds are so common, we tend to overlook or dismiss their potential threat.

You will hear people say that the mold problem has been blown all out of proportion, that mold is more of a nuisance than a threat. Are they right? Is there any proof that mold causes serious diseases? What do the experts say?

Following are just a few quotes from respected institutions that represent the current state of our knowledge and understanding about the health effects of mold.

The official position of the United States Environmental Protection Agency (EPA) has been stated in their guidance documents for mold: **"All molds have the potential to cause health effects.** Molds can produce allergens that can trigger allergic reactions or even asthma attacks in people allergic to mold. Others are known to produce potent toxins and/or irritants." (Source: www.EPA.gov/mold)

According to a 2004 University of Connecticut Health Center report, **"Fungi can cause health problems to both humans and animals** by several different biological mechanisms: infections, allergic or hypersensitivity reactions, irritant reactions, or toxic reactions."

A workplace mold study completed by the Finnish Institute of Occupational Health links adult-onset asthma to workplace mold exposure: **"Our findings suggest that indoor mold problems constitute an important occupational health hazard."** The Finnish workplace mold study estimated that the percentage of adult-onset asthma attributable to workplace mold exposure is 35% (reported in *Environmental Health Perspectives* (EHP), May, 2002).

Papers have appeared in *Systematic and Applied Microbiology*, the *Journal of Environmental Monitoring*, and the *Journal of Occupational and Environmental Medicine* that link effects and impacts on health to mold.

Although there are still debates and ongoing studies, there is no longer any question that mold can cause and aggravate adverse health conditions in humans and animals. It is an established fact.

The good news is that we need not accept as inevitable the chronic or life-threatening diseases caused by mold. While conventional remedies using chlorine bleach and toxic fungicides are often ineffective and can be

more harmful to human health than the mold they are meant to eliminate, we are happy to report that the data we have collected show that essential oils offer humanity a powerful, all-natural means of addressing the problems of mold, toxic mold, and mildew.

The air-sample data presented in this book show that the blend of essential oils used in these case studies not only killed the mold but actually removed mold spores from the air. And there is indirect evidence that the toxins, volatile organic compounds (VOCs), and allergens associated with mold may also be removed by this treatment. Other types of sampling, used in the case studies presented in this book, show that mold growth is effectively eliminated on wood, wallpaper, and drywall when the essential-oil protocol developed by the authors is utilized.

This 100% natural, non-toxic means of dealing with toxic mold is surprisingly effective on multiple mediums, and it appears to offer benefits to both property and health. This is an important application of essential oils that has been overlooked by the general public due to the mistaken idea that essential oils are only useful as perfume or aroma and flavor enhancers.

In this book we review the science developed by universities and other institutions and present twenty original case studies that scientifically document the field data collected. We also include anecdotal evidence, claims, and counterclaims. We separate fact from fiction, urban legend from science and sound evidence. In Part 4 of the book, there is an in-depth statistical analysis of the field data, and a discussion of the meaningful conclusions that can be drawn from the data that was collected as well as identification of additional research to be accomplished.

Case studies presented in this book are based on tests that utilized a proprietary blend of essential oils containing Lemon (*Citrus limon*),

Cinnamon Bark (*Cinnamomum verum*), Clove (*Syzygium aromaticum*), Eucalyptus (*Eucalyptus radiata*), and Rosemary (*Rosmarinus officinalis* CT cineol) therapeutic grade essential oils. The data show that this essential oil blend and the protocol developed by the authors for these studies and described in this book are far more effective in eliminating toxic mold than any other treatment currently available.

The authors recognize the need for, expect, and encourage continued research to determine the validity and efficacy of essential oils for this purpose. We believe good science will provide additional support to document the true worth of therapeutic grade essential oils in eliminating the threats to property and to health posed by mold, mildew, and toxic mold. However, it is our opinion that the general public need not wait for additional studies to be completed. It is our firm, wholehearted belief that the wealth of controlled scientific studies that currently exist (developed during the past two centuries), combined with the scientific data collected in the case studies performed by the authors, along with current and historical anecdotal evidence from around the world, clearly demonstrate that therapeutic grade essential oils are the non-toxic solution to toxic mold.

We began our research with a question for which we truly did not know the answer:

Can essential oils be used to effectively remediate mold in buildings?

We also asked:

Is there a non-toxic solution to toxic mold?

What we discovered was surprising, unexpected, and in some cases life-saving. The data proved to us that the answer to these two questions is: Yes!

Do not take our word for it. Review the data yourself.

We have every confidence that the essential oils we used work and work extremely well because the data prove it. More than that, we have seen lives changed for the better, and our clients have experienced a wide variety of support and benefits to the body by using this treatment option that more than exceeded their hopes and expectations. Examine the data. Test the protocol. See for yourself.

> **Essential oils offer humanity a powerful, all-natural means of addressing the problems of mold, toxic mold and mildew.**

Since starting this work, Ed and/or I have been called "The Mold Man," "The Mold Team," "The Mold Masters," and "The Masters of Eauuu!"

It's funny. We laugh and enjoy it just as much as anyone. Humor brings light to an otherwise dark subject. It makes it easier to talk about and to think about a subject, yet it does not detract from the depth or the importance of the subject.

Mold certainly impacts and can destroy a person's quality of life. The discoveries presented in this book can positively impact the lives and health of millions of people. This discovery is new; yet it is based in knowledge as ancient as the written word, and what we present offers proof that there is a cleaner, greener, more environmentally friendly, non-toxic option for mold cleanup and remediation in buildings. This discovery has far more positive benefits than any other treatment currently available, and the essential oil blend used is so safe that it is approved by the Food and Drug Administration (FDA) as a food supplement. Nothing else on the market for mold cleanup or remediation can say that.

This book has been written for all the people who are tired of suffering from chronic allergies, sinusitis, ear infections, colds, and other maladies that may be caused or exacerbated by mold. It is written for the environmental professional, the remediation professional, the homeowner, the property management company, the developer, and also for the scientist.

Attendees at our seminars and those who have viewed our DVD, "Toxic Mold – A Breakthrough Discovery," have asked that we provide a concise, educational presentation that laypeople can use to inform family members, friends, or a professional doing mold remediation for them about the facts and the discoveries made and presented in this book. That presentation is found in Appendix A.

This book is arranged in four parts that may be read in succession for a thorough understanding of why mold is a concern and how to deal with it effectively. The reader may also choose to read any individual Part, or Chapter, such as "The Case Studies" (Part 3, Chapter 4) or "The Close Protocol" (Part 4), and then return to other parts of the book at a later time.

This book does not replace previous books written on mold, including those by Jeffrey May, Richard Progovitz, Dr. Ritchie Shoemaker, and others (see the bibliography and references at the end of Chapter 1). Neither does it replace books previously written on essential oils and aromatherapy (see the bibliography for a complete listing of reference books used or reviewed for the writing of this book).

What this book does offer is original research about a new discovery for treating mold-infested buildings. This discovery is non-toxic, not drug related, and is an eco-friendly, green solution to the problems associated with mold infestations in buildings and the health challenges posed as a result of exposure to mold. This discovery is currently the only option for mold remediation that may be used long-term, as a preventative, and while people and pets remain inside the building – even within the same room. It is an effective antifungal and sporicidal agent that removes mold spores from the air and also offers beneficial support to the body.

We have done our best to present this information in a reader-friendly format. We pray that this information reaches all who are searching for it, and that each and every one of you enjoys a happier, healthier, more productive, higher quality of life.

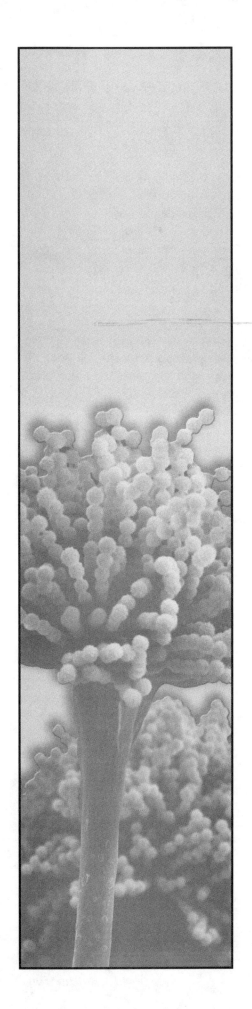

PART ONE

HEALTH AND PROPERTY RISKS ASSOCIATED WITH MOLD

CHAPTER 1

HEALTH IMPACTS OF MOLD EXPOSURE

What Is Mold, and Why Should I Care?

Do you have allergies? How about sinus problems or frequent headaches? According to the American College of Allergy, Asthma & Immunology, "An estimated 40-45 million Americans (15-20% of the population) have some type of allergy."

Researchers at Lawrence Berkeley National Laboratory, Indoor Environment Department, reported in the peer reviewed International Journal of Indoor Environment and Health (June, 2007) that approximately 4.6 million cases of asthma may be attributed to dampness and mold exposure in the home. They further stated that risks of exposure to dampness and mold in schools, offices, and institutional buildings are similar to those in the home, costing $3.5 billion annually.

The Harvard University School of Public Health studied ten thousand homes in the United States and Canada and found that half of them had conditions of water damage and mold associated with a 50 to 100% increase in respiratory symptoms.

What if we reduce our exposure to mold? Would that significantly reduce or eliminate our allergies and other respiratory symptoms?

Everyone has seen, smelled, touched, and probably even tasted mold. There is virtually no place on the planet that is free of mold. Mold thrives everywhere from thirty thousand feet or more in the air to the deepest mine in the earth. It can survive without air, in freezing or boiling hot environments. It grows in the tropics, in the Arctic and Antarctic, and on the highest mountain. It truly thrives and flourishes in the same temperature range that defines the comfort zone for human beings.

Mold can be found in almost every part of our homes: in our kitchens, bedrooms, showers, bathtubs, refrigerators, attics, basements, carpets, pools, backyards, sheds, storage spaces, everywhere. Mold is also found in our offices, in hotels, in schools, and even in hospital clean rooms.

Molds are probably the most successful life form on earth or at least right up there with insects and bacteria. Mold can be green, black, brown, white, pink, yellow, practically any color. And because mold is so ubiquitous, people think it is harmless. We think we know what mold is and that it is not a danger to our health.

Propagated by airborne spores, mold will settle and produce new colonies and many more spores wherever conditions provide sufficient moisture and a source of food. Does that mean arid and desert climates are free of mold? No, it does not. As is seen in the small selection of stories in this book, mold creates problems for people living in the arid parts of the U.S. states of Utah, California, and Montana. Indoor environments, regardless of location, provide conditions that are conducive to mold growth.

All known types of mold are members of a very large family called fungi. An Italian botanist, P. A. Michele, published the first descriptions of fungi in 1729. Fungi include mold, mildew, mushrooms, slimes, yeasts, smuts, and crop rusts. When scientists first started to study and classify these life forms, fungi were thought to be part of the plant kingdom, but they do not quite fit into the common categories of animal, vegetable, or mineral. And as the study of fungi, called mycology, proceeded, it became clear that there were problems with classifying fungi as part of the plant kingdom. For instance, sometimes they behaved more like animals than plants. And there were other characteristics that, in the 1960s, caused scientists to reclassify them into a kingdom of their own, the fungi kingdom. For the first time, fungi were recognized as a unique life form, separate from plants, animals and minerals.

> **Anywhere there is dust, dirt, cellulose, including paper, wood, and cardboard, or any food source, all you need is one mold spore and moisture for mold to grow.**

Compared to most of the plants and animals on Earth today, fungi are also much older life forms. Fossil records show human beings have existed on this planet for eighty to one hundred thousand years, while some fungi have been around for about one billion years. Molds and plants emerged on planet Earth at about the same time, and because plants are made up of cellulose, they are the primary food source for many mold species. If plants had not developed a means to prevent mold from destroying them and consuming them for food, there would probably not be many plants on the planet.

That said, the fungi kingdom includes some very important organisms, in terms of both their ecological and economic roles. Fungi break down dead organic material, and there are plants that could not grow without the symbiotic fungi that inhabit their roots and supply essential nutrients. Other fungi provide drugs such as antibiotics, as well as foods to eat, including mushrooms, truffles and morels, and fungi are also responsible for the bubbles in bread, champagne, and beer. Very adaptable, fungi can change in form and function depending on many factors. For instance, some mold species are generally non-toxic; however, when their space is invaded by other, competing fungi species, then these non-toxic molds can become toxic. They will produce toxins in order to kill or dissuade invaders. Unfortunately, molds also create many problems for humans.

Mold may cause health problems that range from itching eyes, sneezing, and coughing to serious allergic reactions, asthma attacks, bleeding lungs, and even death. What many people do not realize is that mold is probably growing in their homes and workspace right now. Homes and workplaces that are perfectly neat, amazingly clean, where no visible mold is evident, may be hiding super-colonies of mold in ductwork or central heating and air conditioning systems, basements, attics, crawl spaces, under floors, and behind walls.

Anywhere there is dust, dirt, cellulose, including paper, wood, cardboard, or any food source, all you need is one mold spore and moisture for mold to grow.

Moisture can come from leaks, condensation, from a faulty humidifier, or from high levels of humidity in the air.

Some people know they are allergic to mold, yet they may not realize that mold is growing in their homes and workspace. Mold is not visible to us until it creates a super-colony of fruiting structures. Mold spores are microscopic. We cannot see mold spores in the air, even when there are tens of thousands of spores in each cubic meter of the air surrounding us.

> **Mold is not visible until it creates a super-colony of fruiting structures. Even when there are over 50,000 spores in every cubic meter of air surrounding us, we cannot see mold spores.**

Early Warning Signs of Mold Exposure

Do you or anyone in your family have one of more of the following early warning signs of mold exposure?

- € Allergies
- € Cough and Chronic colds
- € Sinusitis
- € Ear infections
- € Fever
- € Dermatitis
- € Headaches
- € General fatigue or malaise
- € Dizziness
- € Flu-like symptoms
- € Nausea
- € Nosebleeds

A 1999 study by the Mayo Clinic found that 96% of the 37 million Americans who suffer from chronic sinusitis symptoms do so because of mold exposure.

Dr. Clive Brown, a medical epidemiologist with the Center for Disease Control (CDC), said in an interview with the staff at *achoo! Review* (www.achooallergy.com/brown-interview.asp): "Over 80% of people with respiratory allergy symptoms are sensitive to fungi, and these sensitive individuals may experience congestion, wheezing, irritated eyes, irritated skin, and difficulty breathing in the presence of mold."

We spend hundreds, thousands, and as a country billions of dollars every year on over-the-counter medications, doctors, and prescription medications to treat the symptoms of the illnesses listed above. We incur billions of dollars annually in lost productivity, and we experience a significantly impaired quality of life.

In a study published in the *American Journal of Managed Care*, Vol. 6, No. 3, March, 2000, a group of scientists reported that at-work productivity losses associated with a diagnosis of allergic rhinitis "were estimated to range from $2.4 billion to $4.6 billion [per year]." Dr. Javed Sheikh, Harvard University Medical School, Division of Allergy and Inflammation, in an article entitled "Rhinitis, Allergic" updated March 2, 2007, says, "Rhinitis is defined as inflammation of the nasal membranes and is characterized by a symptom complex that consists of any combination of the following: sneezing, nasal congestion, nasal itching, and rhinorrhea [runny nose]. The eyes, ears, sinuses, and throat can also be involved . . . The total direct and indirect cost of allergic rhinitis was recently estimated to be $5.3 billion per year." Dr. Sheikh says the causes for allergic rhinitis include exposure to pollens and molds.

Environmental Health Perspectives (EHP), a peer-reviewed journal published by the National

12

Institute of Environmental Health Sciences (NIEHS), March 2, 2005, says, "Exposure to mold and dampness in homes as much as doubles the risk of asthma development in children." And in a March 31, 2005, release, *EHP* says, "Office workers in a northeastern U.S. building that had been damaged by water leaks over a period of years were more than twice as likely to suffer from wheezing or asthma, and over three times more likely to suffer from adult-onset asthma, compared with the general adult population."

Based on the above, it would seem to make sense that if we do our best to eliminate mold in the home and workplace, we can reduce the suffering of millions, save people a lot of money, and significantly improve their quality of life. If exposure to mold only led to an occasional bout with allergies or sinus infection or minor respiratory symptoms, then ignoring mold might be warranted. Unfortunately, mold creates far more serious problems when it comes to our health. The list of illnesses that have been associated with mold exposure is truly staggering.

Gary and Josef's Story – Utah

There is nothing quite so devastating as the loss of a child, a little baby who has his whole life ahead of him, full of promise and love. In November, 2005, Gary was working in the jungles of Ecuador when he received a phone call from his wife, Mary, to come home to the United States as quickly as possible because their youngest son, Josef, was in the hospital with a very high fever and the doctors could not figure out what was wrong.

Gary made the fastest trip out of the jungle ever. Racing to the nearest airport, he caught the first flight home. When he finally reached the hospital, he found his beloved son Josef emaciated, near comatose, unresponsive. The eighteen-month-old had a fever pushing

beyond 104 degrees, and he had been this way for four days.

The doctors told Gary that they were mystified. They had his son on high doses of antibiotics, but blood work showed the boy's white-blood-cell count was soaring. The antibiotics were having no effect. The doctors were running every test they could think of, but nothing explained what was happening or indicated how they should proceed.

Gary had been in Ecuador training in surgical procedures with the world-renowned surgeon, Dr. Edgar Rodas, M.D., F.A.C.S., Dean of the Medical School at the University of Azuay in Ecuador, and donating his time and support to The Cinterandes Foundation. Gary and Dr. Rodas were documenting the effects of therapeutic grade essential oils for pre-op and postoperative care for inclusion in a technical paper they were authoring together. Gary also had a strong background and many years of experience in clinical work. So, when the U.S. doctors said they could not figure out what was causing Josef's illness, Gary brought in his microscope and began doing blood work himself.

The MD in charge of Josef's care brought in forms for Gary to sign, authorizing a spinal tap on Josef, to check for spinal meningitis. When asked if the MD had done any of the noninvasive tests for spinal meningitis, the MD told Gary this was the only test they had not already tried.

Gary asked the MD to check Josef's sensitivity to light, which he knew was a noninvasive way to check for spinal meningitis. He asked the MD to do other non-invasive tests, but the MD felt it would be a waste of time. Gary had already done these non-invasive tests himself and was certain, in his heart, that Josef did not have spinal meningitis. So, to protect his infant son from unnecessary risk, Gary refused to sign the papers for the spinal tap. The MD, having nothing else to offer, said that if the spinal tap would not be allowed, then there was nothing more they could do for the boy.

Gary said, "Then, please, let me take my son home." The MD agreed and released the boy to his mother and father, no doubt expecting that he was sending the child home to die.

Moving their darling son, Josef, back to their multimillion-dollar home in the mountains of Utah, the couple cared for him in their own way, using essential oils and prayer, hoping and trusting that God would guide them and somehow effect a turnaround in Josef's condition. Once home, Gary immediately started Josef on 25,000 units of ascorbic acid (Vitamin C) a day. Still not knowing what was causing Josef's problem, Gary deeply prayed for God's guidance and direction. Then, using a urethra catheter inserted in the small boy's colon, Gary inserted the following essential oils:

> 10cc of Myrrh *(Commiphora myrrha)*
> 2cc of Oregano *(Origanum compactum)*
> 5cc Melaleuca *(Melaleuca alternifolia)*
> 5cc Eucalyptus *(Eucalyptus globulus)*

After two hours, Gary inserted the essential oils into his son's colon again, and within ten minutes of this second rectal insertion, little Josef's fever broke. Gary continued this protocol, and within five days little Josef was up and walking again.

Today, Josef is as healthy and normal as any child his age, full of energy and life. You would never guess he had been so close to death. There are no signs of brain damage, nor motor or sensory damage of any kind. And Josef experienced no sensitivities to the oils.

With Josef rapidly returning to a normal state of health, Gary turned his attention to finding

out what had caused his son to become so ill in the first place. After a great deal of research and more blood work, he learned that Josef had been exposed to mold.

They soon discovered black mold growing in their basement, the place where little Josef had played on the floor for hours every day. Tracing the mold to the wall and a small bubble in the sheetrock, Gary tore out a section and found the spaces behind the sheetrock were heavily infested with black mold.

Gary contacted a specialist in toxic black mold he found in Texas who told Gary that if his child had indeed been exposed to toxic mold, it was a miracle that the boy had lived. After consulting with the mold expert, Gary learned that his home was heavily infested with the deadly, toxic black mold *Stachybotrys*. It was growing in the carpet and behind the walls of his basement.

Photo 1: *Stachybotrys* on Wall, Baseboard, and Carpet.

The couple had the basement walls taken down to the studs. They took out wood, sheetrock, plywood, everything that had any hint of mold infestation or water damage. They found the wood studs were rotten, eaten by the mold, so they went farther, taking the basement down to the footings. They found every leak and potential for moisture and repaired each and every one.

Again, following the inner guidance received during prayer, Gary treated everything in the basement with a blend of essential oils. He painted the concrete and the new wood studs with undiluted essential oil. He sprayed the essential oil a second time before allowing the new sheetrock to be installed. He diffused the essential oil throughout his house on a continuous basis. Only then, after all that had been completed, did he have the basement refinished.

Gary says his experiences in Ecuador using mega doses of essential oils for pre-op and postoperative care of surgical patients prepared him for what he had to do to help his beautiful young son Josef return to health unscathed by this attack of deadly toxic mold. He says he wishes with all his heart that every child could be saved from what Josef went through as a result of his exposure to toxic mold and that every parent could be saved from the feelings of helplessness and hopelessness that inevitably assail any parent watching their small infant child being brought to the brink of death by something that leaves even the best MDs baffled and mystified.

Even with his twenty-five years of experience with essential oils, Gary says that he believes no one fully understands the complexity and miracle of essential oils. He also says, "If you turn to Our Father in faith, even when experiences sent to teach us are painful, He will answer." He thanks God for providing the necessary training and materials and for saving his youngest son's life.
It happened that at the very same time Gary was learning about the amazing ability of essential oils to remove mold from the body, Ed Close was doing research that showed that the essential oils Gary used to eliminate mold in his home did a better job at removing mold

from buildings than anything else currently available. Neither of these researchers knew about the other's efforts. So here, at the same time, in two different parts of the country, two highly skilled, highly trained individuals were learning quite independently of each other about the ability of these God-given essential oils to solve the problems created by toxic mold. And whether by God's guidance, coincidence, serendipity, or scientific research, they both chose the very same blend of essential oils for removing mold from buildings.

Based on Ed Close's research, Gary's home will probably never have a mold problem again, especially if they continue to diffuse the essential oil blend on a regular basis as a preventative measure, and do monthly inspections to assure that no new leaks or sources of moisture create new problems. This story underlines a very important point: Mold spores are not visible to the human eye until they form large, super-colonies of spores and fruiting structures. The mold that was in Gary and Mary's home was mostly hidden from view, and only the bubbles or ripples in the surface of the sheetrock gave a hint that there might be a problem.

We have heard stories that are similar to this one many times. People get sick. They go to their doctor, who cannot find a specific cause for their problems. The doctor prescribes antibiotics, which seem to help some of the time, but often people will go through several different types of antibiotics over several months, and even when the antibiotics seem to help, the patient's problems become chronic and the MD cannot find a source or a cause. MDs are not trained to screen for mold spores or their associated mycotoxins in the blood as a matter of course or to consider that mold may be the cause or contributor to a specific health problem. The scientific research, the knowledge, and understanding of these issues are relatively new. Patients do not bring their home environment with them when they go to their doctor to be tested. The real source of the health problems people are suffering from may go undetected for many years, leading to ever more serious health problems.

Children, especially very young children, do not have highly developed immune systems, and that may be why exposure to toxic mold can be lethal when levels are high. Recent research has linked some cases of Sudden Infant Death Syndrome (SIDS) to toxic mold exposure.

From 1992 to 1999 a cluster of at least sixteen infant deaths in the Cleveland, Ohio, area resulted from acute idiopathic pulmonary hemorrhaging (bleeding lungs). In each and every case, one of the commonalities found was exposure to the mold *Stachybotrys chartarum*, a deadly toxic mold. A clinical study currently being conducted by Case Western Reserve (study start: January, 1999; expected completion: February, 2010) for the NIEHS is collecting samples of secretion, blood, and urine from infants diagnosed with idiopathic pulmonary hemorrhage and analyzing these fluids for fungal spores and mycotoxins. In study details provided by the NIEHS, it states that another 138 cases of acute idiopathic pulmonary hemorrhaging in infants were identified nationwide during the four year period 1995 to 1999 and cites a CDC case-control study that found an association with water-damaged homes and the toxigenic fungus *Stachybotrys chartarum*. While this has not been determined definitively, if you have a mold problem, it is best to deal with it quickly.

Ritchie C. Shoemaker, MD, in his book, *Mold Warriors – Fighting America's Hidden Health Threat*, presents proof from numerous case studies that mold causes very serious illness, even death. He calls us to battle in what he terms a war against mold.

Illnesses Associated with Mold Exposure

The illnesses that have been scientifically linked to exposure to mold are surprising. In alphabetical order, they are:

- allergies
- allergic rhinitis
- asthma
- bleeding lungs
- breathing difficulties
- cancer
- central-nervous-system problems
- chronic coughing
- colds (chronic, frequent, recurring)
- coughing up with blood
- dandruff problems (chronic) that do not go away despite use of antidandruff shampoos
- death
- dermatitis
- diarrhea and other types of digestive difficulties
- earaches and chronic ear infections
- eye and vision problems
- fatigue (chronic, excessive, or continued)
- flu-like symptoms
- general malaise
- hair loss
- headaches
- hearing impairment or hearing loss
- hemorrhagic pneumonitis
- hives
- hypersensitivity pneumonitis
- irritability
- itching (nose, mouth, eyes, throat, skin, or any other area)
- kidney failure
- learning difficulties
- mental dysfunction
- memory loss or memory difficulties
- nausea
- personality changes
- peripheral-nervous-system effects
- redness of the sclera (whites of the eyes)
- runny nose (rhinitis) or thick, green slime coming out of nose or other sinus cavities

- € seizures
- € sinus congestion, sinus problems, and chronic sinusitis
- € skin sores, lacerations, and rashes
- € skin redness
- € sleep disorders
- € sneezing fits
- € sore throat
- € Sudden Infant Death Syndrome (SIDS)
- € tremors (shaking)
- € verbal dysfunction (trouble in speaking)
- € vertigo (feelings of dizziness, lightheadedness, faintness and unsteadiness)
- € vomiting

For more information on health related impacts of mold, see www.fungusdoctor.org.

And just what is it about mold that affects human health? The answer is nearly everything. But two aspects stand out: spores and mycotoxins. Both can be airborne, and both can cause irritation and allergic response upon contact, as well as many other serious symptoms and diseases. Spores can be irritants and cause allergic reactions both on the surface of the skin and inside our bodies. They can grow inside nasal passages, sinus cavities, bronchial passages and lungs, living on a combination of particulate matter, mucus, and tissue. Even dead mold spores can produce irritation, allergic reactions, and other health problems.

Some mold species appear to produce only one toxin, while others are known to produce over one hundred. Some molds also produce compounds called synergizers, which enhance the effects of their toxins. When you smell mold, you are breathing microbial volatile organic compounds (MVOCs) produced by mold. These may or may not be toxic. Scientists have differing opinions as to whether the MVOCs that produce the smells are necessarily toxic. More than five hundred

MVOCs produced by mold species have been identified so far.

The study of toxins produced by mold species is still in its infancy. Currently, hundreds of mycotoxins have been identified, and many more are suspected. It is likely that the number will eventually be in the thousands.

What is the difference between mycotoxins and MVOCs?

Molds produce mycotoxins from other chemicals like polypeptides and amino acids, which they use in metabolism. MVOCs, on the other hand, are derived from alcohols, ketones, and hydrocarbons. Mycotoxins are not volatile and are attached to mold spores. MVOCs are gaseous at room temperature, mix easily with air, and impact our olfactory nerves. (References: www.mold-help.org/fungi.mycotoxins.currentresearch.htm and www.germology.com/mycotoxins.htm).

Photo 2: *Aspergillus* on Plaster.

Mycotoxins produced by the *Aspergillus* species are called aflatoxins. *Stachybotrys chartarum* produces three mycotoxins: Roridin E, Verrucarin J, and Satratoxin H[3], and *Stachybotrys* spores have these toxins on their surfaces, probably to protect the spores in an environment made hostile by other, competing mold species. So, in the case of *Stachybotrys*, it is not only the odiferous MVOCs but inhaled spores that carry toxins into the breathing passages and into the lungs. *Stachybotrys* spores are sticky spores that are not normally found in air samples. So, when an infestation is disturbed by cleaning, remodeling, a child playing on mold-infested carpet, or by simple air movement, these spores pose a known hazard to human health.

Mold – Cause or Aggravation for Disease

It may well be that exposure to toxic mold is the most commonly overlooked underlying cause and/or aggravation of numerous health problems. It is not unusual for people to suffer from debilitating symptoms for years, with doctor after doctor telling them that they can find no cause or labeling the symptoms with catch-all terms like "fibromyalgia" or "chronic fatigue syndrome." Because patients do not bring their environment with them into the doctor's office, and house calls are pretty much a thing of the past, the doctor might be forgiven for not relating a patient's symptoms to a mold-infested home or workplace. One of the primary goals of this book is to help educate the general public about mold and the problems associated with mold.

> **It may well be that exposure to toxic mold is the most commonly overlooked underlying cause and/or aggravation of many health problems.**

In 1986, an outbreak of trichothecene toxicosis was reported in a residence in Chicago, Illinois (Croft et al., University of Pennsylvania Medical School). Over a five-year period, members of the family living there complained of headaches, sore throats, flu-like symptoms, recurring colds, diarrhea, fatigue, dermatitis, and general malaise. These are common symptoms associated with exposure to mold. Spore-trap air sampling revealed high levels of *Stachybotrys chartarum* spores. *Stachybotrys chartarum* was found growing in air ducts and on wood-fiber ceiling materials. Trichothecene, a toxin commonly produced by *Stachybotrys chartarum,* was identified in bulk samples taken from debris found in the air conditioning vents and ceiling material. When the mold-infested materials were removed, and the moisture condensation problem corrected, the family's symptoms, diagnosed as trichothecene toxicosis, gradually disappeared.

Since the paper by Croft et al., there have been numerous reports of *Stachybotrys chartarum* in homes and buildings in North America. One report, (Johanning et al. University of Alabama) focused on the health of office workers in a flooded New York office building with high concentrations of *Stachybotrys chartarum* on gypsum wallboard (commonly called sheetrock). The study concluded. "self-reported health status indicator changes and lower T-lymphocyte proportions and dysfunction as well as some other immunochemistry alterations were associated with onset, intensity and duration of occupational exposure to *Stachybotrys chartarum* combined with other atypical fungi."

A rigorous study of an outbreak of disease in a mold-infested courthouse and office building in Canada (Hodgson et al. Department of Radiation Oncology, Princess Margaret Hospital, Toronto, Ontario), pinpointed mold as the cause. People working in the courthouse offices complained of fatigue, headaches, chest tightness, mucous membrane irritation, and pulmonary disease. Dampness and moisture were persistent problems in the building, and some walls were heavily infested with *Stachybotrys chartarum*, *Aspergillus versicolor,* and *Penicillium* species.

Mycotoxins were identified in moldy ceiling tiles and vinyl wall coverings. Researchers concluded that a mycotoxin-induced effect was a likely cause of the symptoms reported.

In the past ten years, a number of researchers studying schools and office buildings with indoor-air-quality problems have documented definite correlations between mold and the so-called sick building syndrome. Common symptoms in humans exposed to *Stachybotrys chartarum* are rashes – especially in areas subject to perspiration – dermatitis, pain and inflammation of the mucous membranes of the mouth and throat, conjunctivitis, a burning sensation of the eyes and nasal passages, tightness of the chest, coughing, bloody rhinitis, fever, headache, and fatigue – (US Dept. of Health Center for Disease Control).

> **The bottom line is that exposure to toxic mold of any kind may result in serious illness.**

According to the US EPA (For more information see "Children's Health Initiative: Toxic Mold" at www.epa.gov/appcdwww/iemb/child.htm), the principal biology responsible for the health problems in buildings is fungi, not bacteria or viruses. This report says, "One of the most significant technical results from this project is that the effect of relative humidity is indirect and that very small amounts of moisture, well below those commonly cited, will permit growth. The amount of moisture required for fungal growth can vary depending upon the material and the organism. *S. chartarum* requires high levels of moisture (effective relative humidity required for *S. chartarum* growth would be 94%) and cellulose-containing materials for growth."

Although there are many unanswered questions about the effects of *Stachybotrys chartarum* on human health, experience tells us that one should never handle materials contaminated with *Stachybotrys chartarum* without proper protection and using appropriate safety procedures. The accumulation of research data strongly indicates that indoor environments contaminated with *Stachybotrys chartarum* are very unhealthy, especially for children, the elderly, and anyone with a suppressed or compromised immune system. The bottom line is that exposure to toxic mold of any kind may result in serious illness.

Chances are slim that we will ever be able to completely eliminate mold from our environment or our personal space, yet, it is well known that we can eliminate or greatly reduce the incidence of illnesses and discomforts caused or aggravated by mold by eliminating as much of the mold found in a building as is possible. The common way to do this is by treating our home and work environments with antifungal agents. Unfortunately, most antifungal agents contain toxic chemicals or physical characteristics that are themselves harmful to human health, sometimes more harmful than the mold.

We are happy to report that thanks to the rediscovery and production of high-quality, therapeutic-grade essentials oils, there is an effective alternative to the toxic agents we have used for many years that do eliminate mold. Non-toxic to humans, these high-quality essentials oils are known to provide support to the body that results in positive health benefits.

The tests and data collected by the authors prove that there is an all-natural, 100% organic remedy that is remarkably effective in eliminating toxic mold from buildings. Anecdotal evidence from homeowners, construction crews, and office workers indicates that the essential-oil treatment also reduces and in some cases entirely alleviates the negative health effects of mold exposure.

Direct comparison of mold-remediation technologies is difficult due to the differing modes of application and methods for measuring results. For example, fogging with chemical fungicides provides a one-time shock treatment of entire rooms or confined spaces, while ultraviolet (UV) light continuously treats only the air stream in the air-handling system of an office, school, hospital, or home. While fogging a chemical fungicide will impact mold-spore source colonies, there is little or no residual effect to prevent reinfestation. In fact, you would not want residual effects with chemical fungicides, because they are harmful to human health. UV, on the other hand, provides continuous eradication of mold spores in the air stream but does nothing to eliminate the source colonies. UV installation is costly and does not eliminate mold that may enter a building through windows, doors, or any other avenue of ingress that is downstream of the filter.

Why Can't I Just Use Bleach?

Bleach, long recommended for cleaning moldy surfaces, compares poorly with most other mold-eradication methods for the following reasons:

1. Sodium hypochlorite (NaClO), the active ingredient in bleach, is produced by infusing sodium hydroxide with chlorine gas, a deadly gas that killed thousands of soldiers before gas warfare was outlawed.

2. If bleach is accidentally mixed with an acid-containing cleaner (such as many toilet-bowl cleaners, drain cleaners, and even lemon juice or vinegar), then deadly chlorine gas is released.

3. Household bleach contains only 3% to 6% sodium hypochlorite, because it is so caustic and dangerous. The remaining 94% to 97% volume of household bleach is water. Using bleach or a bleach solution actually provides one of the three ingredients necessary for mold growth and allows for the quick rebound of mold growth, usually within 24 hours or less.

4. Bleach kills mold on nonporous surfaces, but not on porous surfaces and does nothing to the spores that are in the air. When the sodium hypochlorite dissipates, usually within a few hours of application, then mold growth can rebound. Application of liquid bleach to porous, mold-infested materials (such as wood, wallpaper, sheetrock, and the grout between bathroom tiles) will bleach the spore colonies and kill mold spores, but not all of the spores will be killed, and even dead spores can cause allergy symptoms.

In Chapter 3 ("Show Me the Science"), and in Chapter 4 ("Case Studies"), research data, information, and case studies are cited and presented that explain and document the comparison of the use of therapeutic grade essential oils with other mold-remediation methods. Chapter 5 ("Questions, Answers, and Conclusions") has objective comparisons of most existing mold-eradication treatments with the essential oil blend used in the case studies presented in this book.

Studies into the impact mold and associated mycotoxins have on health will continue. Misinformation that suggests using bleach, toxic fungicides, and other toxic chemicals will continue. Misunderstandings and misinformation about the effectiveness of UV, ozone, and HEPA filters and misuse of these technologies will continue.

It is such a blessing to know that you can eliminate toxic mold and all the toxic chemicals and dangerous substances commonly used to control it with an all-natural, 100% organic substance. You have in your hands the information about a newly discovered, people- and pet-friendly option for eliminating mold and the health challenges that may be caused by mold.

References:

American Academy of Pediatrics (AAP), Committee on Environmental Health, "Toxic effects of indoor molds," 1998.

American Academy of Pediatrics (AAP), Policy Statement, "Spectrum of noninfectious health effects from molds," December 1, 2006.

California Department of Health Services, Environmental Health Investigations Branch, "Health effects of toxin-producing indoor molds in California," April, 1998.

Dearborn, Dorr, PhD, MD, Rainbow Babies and Childrens Hospital, Case Western Reserve University School of Medicine, Department of Pediactrics, Division of Pediatric Pulmonology. "Cleveland cluster of infant pulmonary hemorrhage: A Stachybotrys connection?" 2000.

Fujimura, M., Y. Ishiura, K. Kasahara, T. Amemiya, S. Myou, Y. Hayashi, Y. Watanabe, E. Takazakura, A. Nonomura, and T. Matsuda. 1998. Necrotizing bronchial aspergillosis as a cause of hemoptysis in sarcoidosis. *Am J Med Sci.* 315:56-58.

May, Jeffrey C., *My House Is Killing Me*, The Johns Hopkins University Press, 2001.

May, Jeffrey C. and Connie L., *The Mold Survival Guide for Your Home and for Your Health*, The Johns Hopkins University Press, 2004.

Mudari, D., and Fisk, W.J., "Public health and economic impact of dampness and mold," *Indoor Air*, the International Journal of Indoor Environment and Health, Vol. 17, issue 3, page 226-235, June, 2007.

Progovitz, Richard F., *Black Mold, Your Health and Your Home*, Forager Press, 2003.

Severo, L. C., G. R. Geyer, and N. S. Porto. 1990. Pulmonary *Aspergillus* intracavitary colonization (PAIC). *Mycopathologia*, 112:93-104.

Shoemaker, MD, Ritchie C., *Mold Warriors*, Gateway Press, 2005.

Stevens, D. A., V. L. Kan, M. A. Judson, V. A. Morrison, S. Dummer, D. W. Denning, J. E. Bennett, T. J. Walsh, T. F. Patterson, and G. A. Pankey. 2000. Practice guidelines for diseases caused by Aspergillus. *Clin. Infect. Dis.* 30:696-709.

Vance, BA, Paula H. and Weissfeld, PhD, Alice S., *Clinical Microbiology Newsletter*, Vol. 29, No. 10, 73-76, "The controversies surrounding sick building syndrome," May 15, 2007.

CHAPTER 2

MOLD AND PROPERTY —
CAREFUL!   IT COULD BE A MONEY PIT

A couple moves into their dream home, a twenty-two room, 11,500-square-foot mansion on seventy-two acres. They are thrilled! It is beautiful beyond their wildest hopes and expectations.

What homeowners have not felt the joy of closing on a new home? It is a giant step for them and their family. Unfortunately for this particular couple, their dream home became a nightmare, and the story of their mold-infested Texas mansion became famous, appearing in newspapers, on radio and television programs across the United States. An article, dated December 5, 1999, still appears on the USA Weekend website. Here is an excerpt:

> It started with a series of leaks. Within a year, Melinda Ballard's…Texas dream home was quarantined; her 3-year-old son, Reese, was on daily medication to treat scarred, asthmatic lungs; her husband, Ron Allison, had lost his memory along with his job; and the family was living out of suitcases and locked in a seemingly endless battle with their insurance company. The problem? Household mold.

They had *Stachybotrys*, an especially toxic mold.

It is hard for most people to believe that a little water damage can lead to so much misery, but it happens more often than you might realize. Here was a million-dollar mansion that looked beautiful, had no visible mold, no smell, and yet it cost the owners far more than the price of remediating the mold. And that remediation cost was significant, in the hundreds of thousands of dollars. Yet compared to the cost to their lives, their quality of life, the remediation cost was insignificant.

Do you know someone who has lost their home or a building they own to mold? We learn of new cases almost every day where buildings have been burned, quarantined, or abandoned due to the presence of mold.

Photo 3: *Stachybotrys* Growing on Ceiling Insulation.

Lauri's Story - California

A friend of ours in California called us when her office coworkers started complaining about nausea and headaches. There had been heavy rains the previous weekend in the Los Angeles area, and in the building where they worked there had been several water leaks. Black mold was growing up one of the walls in her area. She asked us what to do.

We suggested contacting a local professional, getting mold sampling done, and then diffusing a blend of essential oils. Even though we suggested they wait until after samples had been taken, they started the diffusers within an hour of our conversation, because our friend had them at her home, and because we had said it would help the people working there. Her coworkers noticed an almost immediate difference in the way they felt once the diffusers started. Work continued. Everyone went about their daily activities with little or no problem.

Later, when the sampling and analyses were completed, no mold was detected. In addition, one of the people working in the office, Lauri – who had become progressively more ill and debilitated over the past few years – noticed that she felt better when she came to work. They placed the diffuser on Lauri's desk and ran it with the essential oils all day, every day. Lauri noticed she continued to feel better whenever she was at work. So Lauri called and asked us why that would happen. We asked, "Do you have mold in your home?" She did not know. There was no visible mold, at least nothing obvious.

Lauri did not have much money. She could barely afford to feed herself after paying her rent. She had been ill for so long, she only worked part-time, and the quality of her work had suffered for years due to her health problems, so her wages were below the poverty level, and she lived in one of the

highest cost-of-living areas in the United States. Taking these factors into consideration, we suggested she contact the Department of Health to see if they would do mold sampling for her at no charge. They did. They found *Stachybotrys chartarum* at such high levels that they quarantined her apartment. She was not allowed to go back inside for anything. She lost everything she owned. She had to move. She could not take anything from her apartment, not a single item. She had to buy new clothes and burn the ones she was wearing. It is hard for most people to imagine the devastation of such an occurrence.

Today, over a year later, Lauri's life is different. She has regained her health. Her work has improved. She is functioning at nearly the same levels as she did before exposure to the toxic mold. Lauri credits the essential oils for, first, giving her hope and then turning her life from a downward spiral into an upward, hope-filled opportunity.

Lauri says she was fortunate. Her church family helped her. They provided money for a diffuser and essential oils for her to diffuse in her new living quarters. Her office continued to supply essential oils and a diffuser while she was at work. Amazingly, this remarkable woman is thankful for what happened to her. She says it was a blessing that gave her back her life. "Things," Lauri says, "can be replaced, but your health affects everything in your life, and when you are miserable, things offer no comfort."

The value of this newly discovered treatment option is that even people with modest means have something that can help them if they or their property have suffered as a result of exposure to mold. And this same option can be used to prevent mold without fear of poisoning themselves, their family, or their pets.

The Cost of Mold Remediation and Who Pays It

The cost for cleaning up and remediating mold, using the old, standard methods employed by most mold-remediation companies, is high. In 2001, insurance companies paid $1.3 billion dollars in claims for damages to property by mold. In 2002, they paid more than $3 billion in mold-related property damage claims. In 2003, the insurance companies quietly eliminated mold-related coverage. Surprisingly, this issue was not contested by consumers.

In an article entitled: "Is Mold Toxic to Insurers?" (Reference: http://library.findlaw.com/2004/Oct/27/133605.html), author Curtis P. Cheyney says:

> In a pro-active move, insurers have taken measures to bar coverage for mold completely, or limit their exposure for mold-related claims. Some insurers have left certain markets altogether.

Recently, some states in the U.S. have begun introducing legislation that requires that insurance companies doing property and casualty business in their state provide limited coverage for mold-related claims. And many insurance companies continue to pay to repair damage caused by water leaks, freezing, and flooding (if covered in the property owner's policy), but they will not pay to remediate mold. Worse, if you file a claim for mold cleanup or mold remediation, if you even mention mold in a claim or in discussions with your insurance agent, your name and the address of your property will be entered into a database and that information will follow your property for the rest of its existence, even after the mold problem has been remediated. This will negatively impact the value of that property and its salability forever. In some cases, no insurance company will provide insurance for that property ever again.

That may seem like a bold statement. And it would be rash, if it were not also true.

In the article "Insurance Database Blacklists Homeowners Who File Claims," Dan McSwain, a reporter and managing editor for the *North County Times* in San Diego County, California confirms the system's existence.

> Insurance companies say they are simply responding to a flood of claims, particularly a sharp increase in damage from leaky pipes and resulting mold infestation. The effort to cut risk has caused reliance on an obscure computer system that tracks histories of claims.

The "obscure computer system" McSwain refers to is known as the Comprehensive Loss Underwriting Exchange, or the CLUE database. CLUE compiles information on insurance policyholders, U.S. insurance claims, and claims inquiries filed during the past ten years.

According to some sources, insurance agencies across the country access this database to decide whom they will insure, and as maddening as it may seem, coverage may be denied individuals for prior phone call inquiries made to their insurance agents – even if the call did not result in a claim being filed. McSwain reports, "...both regulators and [insurance] industry executives suggest that owners fix even major problems themselves if they can afford it, and avoid notifying their insurance agents about the repairs."

If you are interested in obtaining a CLUE report for residential property you currently own, you can do that by visiting the website www.choicetrust.com. There is a fee for the report. According to the Choice Trust website, the CLUE Personal Property report provides the "five year history of losses associated with an individual and his/her personal property." As an aid to home sellers, they offer a Home Seller's Disclosure Report, which includes any listed loss, the date of the loss, loss type, status, amount paid, policy type, and the name of the insurance company.

> The adage, "If you cannot see any mold or smell any mold, then you do not have a problem" is faulty reasoning, because mold spores are not visible to the human eye, and mold cannot be smelled until it releases volatile organic compounds (VOCs), which only occurs during certain parts of the mold's life cycle.

Homebuyers concerned about possible previous insurance losses experienced at a property can require sellers to provide a CLUE report as a contingency to their purchase offer. By requesting the CLUE report before closing on a home, or for any residential property you have recently purchased, you will also see the five-year history for the property and the owner of the property during that same five-year period, whether you owned the property during that time or not. Choice Trust says mistakes in the CLUE report are rare. However, a homeowner can challenge the accuracy of specific information, and it might be worthwhile knowing what is in that report before you put your home on the market because it will affect your ability and a prospective buyer's ability to secure insurance as well as financing.

Know what is covered by your homeowner's policy so that you can accurately report legitimate claims to your insurer. Remember, under a standard home insurance policy, mold damage – just like damage from rust and dry rot – is now specifically *excluded* from coverage.

What this means is that when a property has a mold problem, the property owner will probably bear the cost to clean it up alone.

This is not always the case, but it is generally true. Numerous lawsuits are brought every year against builders, landlords and property owners, and contractors when there is evidence of wrongful actions on their parts. However, litigation is a long-term affair, and it can take ten to twenty years to ever see a single penny even if you win the case.

Fortunately, there is now another option to eliminate mold that can be employed at a relatively low cost, and it can be used to keep mold from coming back. Even better, the option presented in this book can be used as a preventative so that mold does not become a problem. When compared to standard mold-remediation practices, the discovery made by the authors can reduce costs exponentially.

The Remediator's Costly Wisdom

The standard remediator's wisdom says, "If you have water damage, or if you can see and smell mold, you already know you have a problem. You are going to have to tear out all the damaged materials, and wash everything down with a fungicide anyway. So, why waste money doing sampling?"

This standard "remediator's wisdom" is wrong more often than it is right because it is based on the incorrect assumption that all mold infestations are alike. They are NOT!

More than 100,000 species of mold have been identified, and only a hundred or so are toxic.

While research has linked many serious health problems to toxic mold, and more links are being discovered every day, a cleanup method that is safe for some mold infestations can be disastrous for others. And if you do not know whether you are dealing with a toxic mold or a benign mold, then you must treat every mold infestation as if it was the most toxic type of mold found on the planet. That can be extremely costly.

Conversely, if you treat the mold as if it is benign, then you may create health problems for yourself or spread the problem throughout the building. And if the materials happen to be infested with a toxic mold species, then your actions could create a far more serious and costly problem to clean up.

Let's look at this remove-and-replace approach with a little clarity. Suppose you had aches and pains from the top of your head to the tips of your toes and everywhere in between. You go to your doctor and say: "Doc, I hurt all over. My head hurts, my feet hurt, my knees, my hands, and my stomach. I've got chest pains. I get gas and heartburn frequently. I seem to be out of breath way too easily, and my lungs hurt every time I take a deep breath." To which the doctor responds: "No problem. We don't need to do any tests. We'll just remove everything that hurts."

Would you think that doctor had your best interests at heart? Would you trust him or her? No, probably not. Yet, this is exactly what many people doing mold remediation suggest be done in buildings where mold is found.

Think of it this way: How can you possibly know what to do if you do not know what the problem is? And how can you know what the problem is if you do not do proper sampling and testing?

Following are a few examples from personal experience that will help illustrate the importance and benefits of proper sampling.

1. At one site, the sources of mold spores in a hospital clean room were found through careful use of three different sampling techniques. This facility had expensive HEPA filters and UV protection for the building's air-handling system and used EPA-approved cleaners that are specifically designated for hospital clean rooms, yet this hospital had a mold problem. Without proper sampling, the source of the mold would not have been identified and the long-term liability and risk to the hospital would have been in the millions of dollars.

2. At numerous sites, including residences and commercial buildings, sampling revealed totally different mold species in different rooms. In all cases, considerable money was saved because less-drastic and less-expensive cleanup and remediation methods could be used in the rooms that did not contain toxic-mold growth.

3. At two sites, one a residence and the other a commercial property, the expense of tearing out and replacing many square yards of drywall was avoided when proper sampling showed that the mold infestations were only on the surfaces of the walls and floor. Essential oils were diffused, and the household cleaner with the essential oil blend was used to clean the surfaces. Bulk, tape lift, and air-vacuum spore-trap samples taken after treating with the essential oils used by the authors showed that the mold had been eliminated (for additional information about sampling techniques, see Chapter 7, "Sampling Methods, Their Advantages and Disadvantages").

The most important reason for inspection and sampling by a qualified professional

26

is to ascertain whether you have toxic mold.

The second most important reason for sampling by a qualified professional is to determine whether your remediation efforts have been successful.

One of the most serious mistakes property owners with mold problems make is when the mold is assumed to be benign, and is treated with bleach or other toxic chemicals without using the proper protective clothing and equipment. The least serious consequence of this action includes short-term illness with irritation to eyes, ears, skin, and the respiratory system. In some cases, exposure has resulted in hospitalization and/or long-term health problems.

The "tear it all out" approach can also mask serious problems. After all, if the infested materials have been recently removed and everything refinished, you cannot tell just by walking into that building that there is a mold problem.

The adage, "If you cannot see any mold or smell any mold, then you do not have a problem" is also faulty reasoning, because mold spores are not visible to the human eye. Mold spores are microscopic. You cannot see mold unless or until it has formed super-colonies of fruiting structures. There can be 10,000 spores per cubic meter of air in a room, and you would not know mold was there unless you are sensitive or allergic to mold spores. Plus, mold can only be smelled when it releases volatile organic compounds (VOCs), and that only occurs during certain times in the mold's life cycle. The best and possibly the ONLY way to detect mold that is not visible or apparent by smell is through sampling.

Now, let's assume that someone has a mold problem in their home and they follow the

standard "remediator's wisdom" and have everything torn out and replaced that has any visible mold, staining, or water damage. As soon as the house is remodeled, it is put up for sale. The home looks new and perfect.

A buyer comes along and falls in love with the home. It's clean. It looks new. The buyer asks for a mold inspection, and the realtor says, "Sure. No problem at all, if you think that is really necessary. Of course, it will cost extra. It may mean a longer escrow, and if there is another offer that does not ask for that inspection, it could mean you lose this house."

The home looks great. No mold is visible anywhere. There is no musty odor. A buyer, in this case, might easily decide that a mold inspection is probably just a waste of time and money. That could be a serious mistake.

That is how easily someone can buy a building or a home that looks perfect, a dream come true, yet a few weeks or months later they find they are chronically ill with respiratory infections, headaches, and sometimes more serious health problems. They do not suspect mold or even consider it as a possible cause for their health problems. Yet, the illness continues, and then, one day, they find the first sign of mold growing. Or maybe a neighbor tells them how run-down the house was before the renovations. Or maybe a friend suggests they might be suffering from mold exposure. They call in someone who uses the old, standard "remediator's wisdom" and learn mold is growing between the interior and exterior walls, in the attic, and in the crawl space. They have purchased a "money pit" that is the standard remediator's pot of gold. In our experience, an average cost for mold remediation using the current, standard tear-it-all-out technique is in the $10,000 range for a single room up to 1,000 square feet of dwelling space.

Chapter 2

Without having an inspection and sampling done prior to buying the house, there is no way to prove the condition existed prior to purchase, and that leaves the buyer with no recourse to the previous owner or the real estate agent. To protect yourself, your employees, and your family, you would be well advised to have any building inspected for mold before taking possession or moving into it. And when you know that, as the owner, you are responsible and will bear any and all mold remediation and cleanup costs alone, then your best way to reduce risk is to be certain you are not purchasing a mold-infested money pit.

Mold inspections and sampling can cost as little as $350 to $1,000, depending on the size and location of the building. However, if you buy a building that has mold problems, it can easily cost many thousands of dollars to remediate the mold. You also risk having the building condemned, losing it and everything inside it, and in the worst case scenario, you could lose the building, all your possessions, and your health and quality of life.

The best news is that using the protocol developed and presented in this book will prevent and eliminate most molds. Based on the field data and research conducted to date, it not only destroys mold, it actually removes the spores from the air.

It is strongly recommended that before using the essential-oil protocol, you have sampling done by a qualified professional to help determine the best approach for your specific situation. Because mold is present everywhere, the self-test kits available on the Internet and at home-improvement stores will always show mold present in a building. They do not provide accurate or sufficient information about the amount of mold or the types of mold present. We urge that you not waste your money and your time on these kits. Chapter 5 has more on this.

Construction and Mold

Often a mold problem starts with building construction. In fact, preventing mold from developing in lumber that lays waiting at the lumberyard in all sorts of weather conditions is a major area of mold-related research. That is only one of many areas in the construction process that contributes to the increasing problem of mold in modern buildings.

Many times, serious mold problems can be avoided if building contractors are willing or able to change certain common practices. We are not suggesting that contractors are deliberately doing things that cause mold problems; just that in the heat of the battle to get permits, comply with federal, state, and local regulations and ordinances, get the right construction materials on site as needed, and meet scheduling goals and deadlines, mold is not the uppermost thing on the builder's mind.

In humid climates, one of the leading causes of mold problems in new houses is that contractors do not allow wet materials to dry before incorporating them into the building. For example, if a home is being built in the Midwest, groundbreaking may be postponed until spring to avoid frozen ground and the problems associated with that. Suppose sheetrock is ordered in May, but lumber procurement and scheduling delay installation of the sheetrock until July. Thunderstorms roll in, and the sheetrock, stacked on site, is covered with a plastic tarp. The sun comes out the next day. The standing studs are wet, and the sheetrock, under the plastic, is damp. But it is a beautiful day, and pressures from the property owner and other waiting jobs prompt the builder to forge ahead. The sheetrock is put in place. The house goes up rapidly. And moisture is sealed inside the walls right along with the ever-present mold spores.

The family moves in, and within a few weeks or months, health problems like headaches, chronic cold or flu symptoms, and memory loss begin to show up. The unsuspecting owners, who keep their home in immaculate condition, never even realize that mold is the cause of their problems. Why? Because they do not see mold anywhere. They do not smell mold. They do not know the walls were enclosed with damp studs and sheetrock. They have no idea there is mold growing behind their walls.

Photo 4: Heavy Growth of *Cladosporium* on Sheetrock in an Office Building.

In some cases, mold spores may lie dormant until a plumbing problem, a roof leak, or some other additional source of moisture causes a mold infestation to blossom forth, which might occur months or even years after the house is built.

If you own your own a home, or are a renter, you can avoid mold becoming a problem in your living space by always repairing leaks quickly, cleaning regularly, and using the protocol developed (presented in Chapter 7) as a preventative. Diffuse the essential oils used in the field tests in your home on a regular basis to protect your home, your family and your pets.

Another common construction mistake is the failure to install moisture barriers and adequate drainage under houses with crawl spaces, or earth-contact. Older homes with cinderblock basement walls or concrete walls that become porous with age are virtual mold incubators. Construction mistakes that result in moisture intrusion and mold growth may not show up for years.

Normally healthy, young to middle-aged adults may not notice any health effects from mold. That does not mean they are not affected by mold, but often they do not make the connection between mold and chronic cold, flu, or other respiratory conditions. Many times they only begin to think that something may be wrong with their home if the chronic symptoms clear up when they are away from home for a period of two weeks or more and then reappear as soon as they return home.

Underestimating or ignoring a mold problem often leads to much more severe, long-term problems. For example, after living in their home for a few years, a young couple decides to remodel. So, they hire a remodeling contractor. In the process of removing a wall, the contractor finds black mold has infested the wall and advises the couple to hire a professional mold expert to evaluate and remediate the problem before proceeding with the remodeling project.

Fearing a quantum jump in costs, the homeowners say: "O.K., we'll get that taken care of and call you when it's done." Yet, prompted by economic concerns, they do not contact a professional. Instead, they clean up the mold themselves, using bleach. Bleach treats the surface, and it appears that the mold has been dealt with, but this quick-fix decision will not last. Within a few weeks, after the remodeling is completed, the water left in the porous materials where they used the bleach will provide the necessary conditions for the mold to return and create a more costly and far more damaging problem.

Another common mistake in the construction of office buildings, banks, and even hospitals occurs when insufficient consideration is given to negative air and vapor pressures, air flow, and vapor barriers when designing buildings and HVAC systems. Starting in about the mid-1970s, commercial building architects and contractors began to provide economic savings through simple, sealed box-like design and greater heating and cooling efficiency. Air-handling systems were placed on flat roofs to save space, and windows were sealed shut to create a functionally closed air-handling system. Most often, the outer walls of these buildings were stucco, rock, or brick veneers. These veneers are porous enough to absorb moisture from rain and humidity. If no moisture barriers with vertical drains exist, either because they were never installed or because they are plugged up, and the first real moisture barrier is vinyl wallpaper or paint on the inside walls, you have created a very efficient mold incubator. It is not uncommon to peel back a corner of vinyl wallpaper and find black mold growing there. However, the lack of visible mold, even behind vinyl wallpaper, does not in any way mean a room is mold free.

Conscientious building contractors will not knowingly use wet or moldy materials. However, mold is not something that contractors are likely to think about until they actually have a mold problem. The construction practices that lead to mold problems usually arise from one of two errors in thinking during the planning stages:

1. Failure to think about mold as a construction problem

2. Underestimation of the seriousness of mold infestation

Clearly, a change is needed in construction thinking so that errors or omissions that lead to excess moisture and mold infestation can be avoided.

Rehabilitating Mold-Infested Property

Until the discovery that essential oils offered a better option for mold cleanup and remediation, there was little hope of rehabilitating a toxic mold-infested property. About the only option open to property owners was extensive and costly remodeling, total demolition, burning, or abandonment.

We have heard of many properties in various parts of the country that, due to mold infestation, have homes that cost several hundreds of thousands of dollars to build yet are no longer worth any more than the value of the land that they occupy. Recently, we heard about a home in central Illinois where the owners had paid nearly $300,000 for the property. The home was less than two years old, and due to a toxic-mold infestation, the property has been appraised at $46,000, the value of the lot.

That huge loss could have been avoided. The owners have filed a lawsuit against the builder. This type of situation need not occur. A new option is available.

Cherie's Story - Montana

Cherie and her husband were looking for a vacation home in the Montana mountains. The ski-resort town they chose had many properties available, but all were pricey. Cherie was looking for a bargain and she finally found a condo that was priced well below other properties of the same type in the area. When she asked to see the property, the realtor told her, straight out, that she would have to wear a hazmat suit just to go inside because the property had not been occupied for several years, and the pipes had burst, and there was a toxic-mold infestation from the water damage.

Cherie had been using essential oils for many years. She knew from first-hand experience that essential oils could take care of mold problems, so she said she still wanted to see the property. The realtor would not take her there. Cherie decided to go alone, decked out in her personal protective equipment, or "moon suit." In spite of her realtor's objections, Cherie decided to buy the property. After going through six different lenders and having all of them turn her down for a mortgage on the property, Cherie started asking why. They said because the condo was known to have a mold problem, and they were not willing to make a loan on toxic mold-infested property.

Finding another way to acquire the property, Cherie began diffusing essential oils at the site. She used Oregano (*Origanum compactum*), a combination of Melaleuca oils, and the essential oils used in the protocols in this book. Afterward, she used the household cleaner with the essential oil blend, and she cleaned up all the visible mold she could find. Then she had a professional come in and do appropriate sampling to determine if there was any mold present. The samples showed no mold.

Cherie was thrilled, but her challenges were not over. Because local people knew there had been a mold problem with this condo, she spent a long time searching for a contractor who was willing to do any remodeling work, even though she had samples showing the condo had no mold in it now. The contractors knew that all the standard treatments were only temporary fixes, and they could not believe that the mold was no longer a problem.

Being a persistent lady, Cherie finally found a contractor who worked with her. After the cleanup and remodeling efforts were complete, Cherie had sampling done again to show the mold problem was completely eliminated and had not returned. That formerly mold-infested property is now worth more than three times what she has invested in it. She rehabilitated what was a condemned property.

That is the power of essential oils when used correctly, the power offered by this new treatment option.

Mold in Vehicles

"The rash started on his ankles and it took us months to realize that he had a fungal infection. We tried everything to get rid of the rash. It would go away for a short time, then come back again. After hearing Dr. Close speak about mold and watching your video, we found mold growing in the carpet of his SUV. Thanks to your information, we used the essential oils products and got rid of the mold. Now, his fungal infection is gone too." Les and Sharon told us their story in January, 2007, less than one year after we did the first program to present the data we had about essential oils effectiveness against mold.

Water-damaged vehicles, flooded vehicles, and even vehicles that have been cleaned but the carpets have not been dried properly can all be infested with mold, sometimes toxic mold. Our cars, trucks, SUVs, boats, horse trailers, all vehicles provide an opportune environment for mold growth and potentially intense exposure to mold.

There are places in the United States and around the world that are flooded every year. Some flooded to the extent that vehicles are completely inundated with water. Research shows that many water-damaged vehicles end up being sold in the used car market. You can protect yourself from buying a vehicle that may have been flooded or water-damaged by following a few simple precautions. Look for the following indicators that a vehicle may have been flooded:

- Mildew, debris, and silts in places where they would not normally be found, such as carpeting, in the trunk, or around the engine compartments.
- Rust on screws and other metal parts
- Water stains, faded upholstery; and discoloration of seat belts, door panels and flotation devices.
- Dampness in the floor and carpeting.
- Moisture on the inside of the instrument panel.
- A moldy odor or an intense smell of Lysol, bleach, or deodorizer that may be used to cover an odor problem.

According to the National Insurance Crime Bureau (NICB), there are unscrupulous salvage operators and dealers that often try to conceal the fact that the vehicles they are selling have been damaged by water and mold. Consumers who suspect a car has been flooded can access information on any car's vehicle identification number (VIN) by completing a free search of the "Flood Vehicle Database" at http://www.nicb.org.

To avoid inadvertently purchasing a flood-damaged car, it is important that you only buy a used car from a reputable dealer, have a certified mechanic look for flood damage, and check the car's VIN number.

Things To Remember

Here are a few things to remember:

1. It does not matter whether you have an old building, a new building, cheap construction or a multimillion-dollar building, you can still have a mold problem if there is sufficient moisture and a source of food.

2. It does not matter whether you live in a humid climate or an arid climate, you can still have mold problems that can damage your property values and your health.

3. You cannot see mold until super-colonies of fruiting structures are formed. You can have a mold problem even when you do not see or smell mold.

4. You cannot smell mold until it enters the portion of its life cycle that produces microbial volatile organic compounds (MVOCs).

5. If you have mold in property you own, you will probably bear the cost for cleanup and remediation alone.

6. The only way to know whether you have mold is by doing appropriate sampling and analysis. A mold inspection and sampling can cost as little as $350.

7. Your best option for dealing with mold is to prevent mold from ever becoming a problem.

For more information on preventing mold, see Chapter 5.

Checklist: Warning Signs of Mold

Check all that apply to your home or building.

- ☐ Increased allergy or respiratory problems for humans or pets.
- ☐ Chronic colds, headaches, or flu-like symptoms.
- ☐ Musty odors, or an earthy smell.
- ☐ Water stains, discoloration, or visible mold growth on any of the following:

windows, floors, carpets, walls,
wallboards, wallpaper, sheetrock, vents,
ceiling, baseboards, tile flooring or
counter tops, windowsills.

- [ ] Mold and/or algae growing on stucco, paint, brick, vinyl or other types of siding on the exterior of the building.
- [ ] Indentations, bubbling, cracked or peeling paint or wallpaper.
- [ ] High humidity.
- [ ] Areas of standing water.
- [ ] Condensation on floors, walls, or windowsills.
- [ ] Warped wood.
- [ ] Water leaks.
- [ ] Flooding.
- [ ] Leaky roof.
- [ ] Leaking pipes.
- [ ] Leaks around windows, fogged windows indicating an improper seal and possible damage to the surrounding window frame and drywall.
- [ ] Blocked gutters.
- [ ] Poor ventilation.
- [ ] Loose drywall tape.
- [ ] Damp basement.
- [ ] Damp crawl space.
- [ ] Mold or discoloration on the patio concrete, in hot tubs, pools, and on decking.
- [ ] Humidifier(s) used in space.
- [ ] House plants.
- [ ] Box, file, and paper storage in basement, attic, or other areas with higher than 40% relative humidity.
- [ ] Areas with poor ventilation.
- [ ] High temperature variation.
- [ ] Building closed for vacations.
- [ ] Building or individual units in the building have been vacant.
- [ ] Vents and Ductwork for heating or air conditioning
- [ ] 10 degrees (or more) fluctuation between daily high and low temperatures

PART TWO

THE SCIENCE

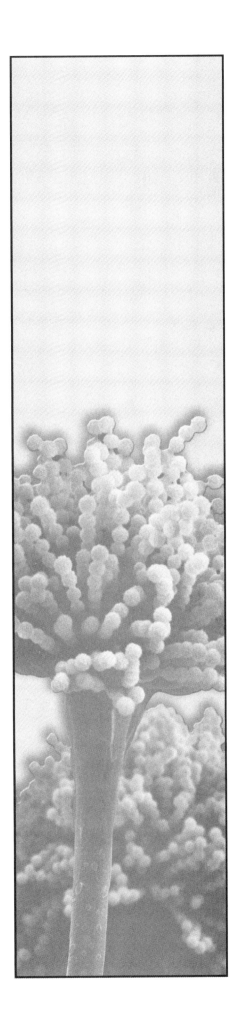

CHAPTER 3

SHOW ME THE SCIENCE

Suggesting that a professional engineer, with more than forty years experience in the environmental field, use aromatherapy to deal with toxic mold is like asking a professional baseball player to go to bat in the World Series with a nerf ball and a sponge bat. Dr. Ed Close was skeptical, hesitant, and not interested in suggesting that his clients even consider using this approach.

Essential oils were fluff and puff, massage and spa, health and well-being, but not serious science, not standard engineering practice. He had a reputation to uphold and, more than that, a responsibility to his clients to provide a standard of service in keeping with his training, experience, and his professional registration. So when Jacqui, a Registered Aromatherapist with more than ten years' experience with essential oils, insisted that if nothing else was working, it was worth a try, Dr. Ed said, "Show me the science."

Internet searches conducted using the combined terms *essential oils* and *antifungal* yielded 152,000 citations on Google and 136,000 citations on Yahoo. At the time this was first researched in 2005, no previous studies were found that assessed the effectiveness of using essential oils for cleaning, removing, or remediating mold in buildings. Numerous studies had proven over and over again that essential oils did possess antifungal properties. Searches conducted in

May, 2007, revealed a single peer-reviewed paper published in 2007 that assessed the antifungal properties of Thyme and Thymol for inhibiting and eliminating fungus in buildings and found them to be very effective.[1]

According to the paper, "Antibacterial and Antifungal Properties of Essential Oils," published in the journal *Current Medicinal Chemistry*:

> The antimicrobial properties of essential oils have been known for many centuries. In recent years (1987-2001), a large number of essential oils and their constituents have been investigated for their antimicrobial properties against some bacteria and fungi in more than 500 reports…This paper gives an overview on the susceptibility of human and food-borne bacteria and fungi towards different essential oils and their constituents. Essential oils of spices and herbs (Thyme, Origanum, Mint, Cinnamon, Salvia and Clove) were found to possess the strongest antimicrobial properties among many tested.[2]

The majority of the scientific studies evaluated the antifungal properties of essential oils and had been conducted for the food industry.

They provide strong evidence for the inhibitory effects of essential oils against various strains of mold.

The Scientific Research

The following are synopses of the scientific papers initially provided to Dr. Ed Close. (In this section of the book, the word *mold* may also be spelled *mould* if the British spelling is used in the original document.) The works included are by no means a complete list of papers available that address the antifungal and antibacterial properties of essential oils.

1. **Evaluation of some essential oils for their toxicity against fungi causing deterioration of stored food commodities.** Applied Environmental Microbiology. 1994 April. [3]

   Synopsis: The essential oil of *Cymbopogon citratus* was found to exhibit fungitoxicity against *Aspergillus flavus*. The fungitoxic potency of the oil remained unaltered for seven months of storage and upon introduction of high doses of the test fungus. It was thermostable in nature with treatment at 5 to 100 degrees C. Findings indicate the essential oil of *Cymbopogon citratus* is an effective inhibitor of storage fungi.

   [**Note** from authors: *Cymbopogon citratus* is commonly known as Lemongrass or Oil Grass. *Aspergillus flavus* is a toxic mold associated with aspergillosis of the lungs and is believed to cause corneal, otomycotic (fungal ear), and nasoorbital infections. This mold is commonly found on corn and peanuts, as well as water-damaged carpets, and is known to produce aflatoxin, a carcinogenic substance. [4-17]]

2. **Inhibitory action of some essential oils and phytochemicals on the growth of various moulds isolated from foods.** Food Science and Technology. [18]

   Synopsis: The aim of this study was to determine the sensitivity profile of mold strains isolated from foods to some essential oils and phytochemicals. The assayed mold strains were: *Fusarium spp., Rhizopus spp., Aspergillus flavus, Aspergillus niger,* and *Penicillium spp.* Among the products that evidenced antimold activity, citral and eugenol showed the lowest minimum inhibitory concentrations for the most of the tested mold strains.

   [**Note** from authors: Citral is an aldehyde found in Marjoram. Eugenol is a phenolic compound found in Clove, Cinnamon, citronella, and Lemon Eucalyptus. [19] *Fusarium spp.* are species of mold known to cause superficial and systemic infections in humans, including keratitis, endocarditis, sinusitis, peritonitis, septic arthritis, and other infections. [20-36] *Rhizopus spp.* are known to cause or contribute to mucocutaneous, rhinocerebral, pulmonary, gastrointestinal, and disseminated diseases. [37-44] *Aspergillus flavus* and *A. niger* are both toxic molds, well known to cause opportunistic infections, allergic states, and toxicoses, including sinusitis, pneumonitis, pulmonary aspergillosis, and other invasive diseases. *A. flavus* produces aflatoxin, a known carcinogen. [4-17] *A. niger* is one of the most common causes of fungal ear infections, may cause temporary hearing loss or damage to the ear canal and tympanic membrane. [45 and 46] *Penicillium spp.* are occasional causes of infections in humans, including keratitis, peritonitis, pneumonia, endocarditis, corneal and urinary tract infections. [24]]

36

3. **Solid- and vapor-phase antimicrobial activities of six essential oils: susceptibility of selected foodborne bacterial and fungal strains.** Journal of Agricultural Food Chemistry, 2005.[47]

   Synopsis: The antimicrobial activity of essential oils (EOs) of Cinnamon, Clove, Basil, Rosemary, Dill, and Ginger was evaluated over a range of concentrations in two types of contact tests (solid and vapor diffusion). The EOs were tested against an array of four Gram-positive bacteria, four gram-negative bacteria, and three fungi (a yeast, *Candida albicans*, and two molds, *Penicillium islandicum* and *Aspergillus flavus*)…Cinnamon and Clove gave the strongest inhibition, followed by Basil and Rosemary. The fungi were the most sensitive microorganisms, followed by the gram-positive bacterial strains. Cinnamon and Clove gave similar results for every microorganism.

4. **Screening for antifungal activity of some essential oils against common spoilage fungi of bakery products.** Food Science and Technology International, 2005.[48]

   Synopsis: The antifungal effect of twenty essential oils against the most important molds in terms of spoilage of bakery products (*Eurotium spp., Aspergillus spp.,* and *Penicillium spp.*) was investigated. Only Cinnamon, Rosemary, Thyme, Bay, and Clove essential oils exhibited some antifungal activity against all isolates. These findings strengthen the possibility of using plant essential oils as an alternative to chemicals to preserve bakery products.

[**Note** from authors: *Eurotium spp.* are the fruiting states of *Aspergillus spp.* and are likely to be present if growth has been long term.]

5. **Antifungal activities of essential oils and their constituents from indigenous Cinnamon leaves against wood decay fungi.** Bioresource Technology, 2004.[49]

   Synopsis: Results from the antifungal tests demonstrated that the essential oils…had strong inhibitory effects…Cinnamaldehyde possessed the strongest antifungal activities compared with the other components. Its antifungal indices against both *Coriolus versicolor* and *Laetiporus sulphureus* were 100%…In addition, comparisons of the antifungal indices of cinnamaldehyde's congeners proved that cinnamaldehyde exhibited the strongest antifungal activities.

[**Note** from the authors: The highest concentrations of Cinnamaldehyde are found in the essential oils of Cassia (85%) and Cinnamon (46%).[19] *Coriolus versicolor* and *Laetiporus sulphureus* are both mushroom fungi typically found on trees. A *congener* is something that belongs to the same class, group, or type, for example, an animal or plant of the same genus.]

6. **Screening for inhibitory activity of essential oils on selected bacteria, fungi, and viruses.** Journal of Essential Oil Research, 1998.[50]

   Synopsis: Forty-five essential oils were tested for their inhibitory effect against bacteria, yeast, molds and two bacteriophage…all oils showed inhibition compared to controls. Cinnamon bark and Tea Tree essential oils showed an inhibitory effect against all the test organisms and phage.

7. **Antibacterial properties of plant essential oils**. International Journal of Food Microbiology, 1987.[51]

Synopsis: Fifty plant essential oils were tested against twenty-five genera of bacteria. The ten most antibacterial oils were: Angelica, Bay, Cinnamon, Clove, Thyme, Almond, Marjoram, Pimento, Geranium, and Lovage.

8. **Influence of some spice essential oils on Aspergillus parasiticus growth and production of aflatoxins in a synthetic medium**. Journal of Food Science, 1989.[52]

Synopsis: The essential oils of Thyme, Cumin, Clove, Caraway, Rosemary, and Sage were tested for antifungal properties. The essential oils completely inhibited growth of fungal mycelium and aflatoxins production.

9. **Inhibition of growth and aflatoxin production in Aspergillus parasiticus by essential oils of selected plant materials**. Journal of Environmental Pathology, Toxicology and Oncology, 1994.[53]

Synopsis: Essential oils of Cinnamon, Thyme, Oregano, Cumin, Curcumin, Ginger, Lemon and Orange were tested on *Aspergillus parasiticus* for antifungal properties. The most inhibitory essential oils were Cinnamon, Thyme, Oregano, and Cumin.

All studies identified thus far that have assessed the antifungal effects of essential oils utilized between one and one-hundred or more single or individual essential-oil species. No studies were found that specifically assessed the antifungal properties or effects of a blend of essential oils.

One study was identified that used a blend of essential oils and determined the antimicrobial effects of the blend against gram-positive and gram-negative bacteria.[54] The blend of essential oils was found to be highly effective against the identified bacteria and this same blend of essential oils (containing Cinnamon, Clove, Lemon, Eucalyptus, and Rosemary) was used in the case studies presented in Part Four.

The Oils Selected for the First Test and Why

Dr. David Stewart reports in his book "The Chemistry of Essential Oils Made Simple"[19] that:

> Ordinarily toxic compounds (like the cineole in Eucalyptus) are rendered harmless and beneficial when mixed with other compounds, as they are in natural oils. Their toxic tendencies are quenched, buffered, and transformed into good by their companions … specific compounds behave differently depending on their company … The effectiveness of essential oils can be enhanced by combining them in special ways … Blends can sometimes be more effective for certain purposes than an oil used as a single because blends usually contain a wider variety of molecular shapes, weights, types, and sizes.

Armed with the knowledge that combining single essential oils into what is called a *blend*, or a *synergy*, may tend to minimize or magnify the effects of the individual oils, and based on the scientific studies above, Jacqui proposed doing three tests in three separate apartments. Ed determined the amount of essential oil required, the length of time for diffusion, the number of diffusers to be used in each space, and other parameters for the physical evaluation of the essential oils' effectiveness [for complete details, see Case Studies 1, 2, and 3, in Chapter 4).

The majority of the scientific studies identified show that the most effective essential oils with a broad range of antifungal action appear to be:

> Cinnamon
> Clove
> Lemongrass
> Oregano
> Thyme

All of these single essential oils are termed "hot" oils, meaning they may cause irritation to skin, eyes and mucous membranes, including nasal passages and lungs, if inhaled for extended periods of time.

Oregano is a strong-smelling, astringent oil that is known to irritate soft tissues and mucous membranes. Oregano oil was rejected because to be effective in a large room, of approximately 800 square feet, the essential oil would be diffused in a space continuously for a substantial length of time. Oregano is not a very appealing aroma, and workers or inhabitants were not likely to tolerate this oil for an extended application. Thyme is also a strong-smelling, astringent oil with many of the same characteristics as Oregano, yet more tolerable in small exposures.

Cinnamon, Clove, and Lemongrass essential oils are also known to be strong smelling and may irritate soft tissues and mucous membranes. However, the smells of these essential oils are much more appealing to most people than either Oregano or Thyme. When combined with other essential oils, they can be quite pleasant.

Knowing this, it was important to look for blends, or synergies, that contained these oils along with other oils that offered the possibility of quenching or buffering the constituents that might cause irritation. Taking this into consideration, Jacqui selected the single essential oil of Thyme (*Thymus vulgaris* CT thymol) and two blends of essential oils.

Essential Oil Blend No. 1 is a proprietary mixture of Citronella (*Cymbopogon nardus*), Lemongrass (*Cymbopogon flexuosus*), Lavandin (*Lavandula x hybrida*), Rosemary (*Rosmarinus officinalis* CT cineol), Melaleuca (*Melaleuca alternifolia* also known as Tea Tree Oil), and Myrtle (*Myrtus communis*).

Essential Oil Blend No. 2 is a proprietary mixture of Clove (*Syzygium aromaticum*), Lemon (*Citrus limon*), Cinnamon Bark (*Cinnamomum verum*), Eucalyptus (*Eucalyptus radiata*), and Rosemary (*Rosmarinus officinalis* CT cineol).

The therapeutic grade essential oils used in the tests conducted by Ed are produced and sold by a company located in Lehi, Utah. Essential Oil Blend No. 1 (EOB1) is sold under the name Purification®.[55] Essential Oil Blend No. 2 (EOB2) is sold under the name Thieves®.[55]

There are several reasons why oils from this company were selected. Having used oils from numerous companies over the years, Jacqui had found that most oils were not high quality, and often were not pure, being either diluted, mislabeled, standardized (meaning they have been manipulated by additions of other species of essential oils or synthetically produced compounds), or adulterated. Because of this, results obtained were often found to be unpredictable. Few companies produce high-quality essential oils for sale in the United States. The following are the reasons essential oils from the company in Utah were selected.

1. Personal experience had proven the essential oils from this company were of extremely high quality and purity.

2. Nicole Stevens, a scientist at Brigham Young University studying essential oils for use in treating cancer, reported at a 2004 conference in Salt Lake City, Utah, that she had conducted tests using essential oils she purchased from numerous companies and had even distilled her own oils from raw materials, but could not equal the quality of essential oils acquired from the company in Utah.

3. There was ample anecdotal evidence that the two blends of essential oils selected were used successfully to eliminate mold without negative side effects and were often reported by users to alleviate or reduce the symptoms associated with exposure to mold.

4. A supply of the oils and the household cleaner containing Blend No. 2 were on hand, therefore, there was no waiting time. The test could commence immediately.

Additionally, the following email from a woman named Deb Bowman suggested that these products could potentially provide a long-term solution to mold.

> Every fall, we get mold growing on the siding of our house, so last October [2004], I cleaned a 4 ft. x 4 ft. area with the Household Cleaner [containing Blend No. 2]. I sprayed it on, scrubbed it with a brush, waited five minutes, and sprayed it down with the hose. I showed my husband how clean it was, but he still chose to clean the rest of the side of the house with bleach. This October [2005], I looked at that side of the house. Again, the seasonal mold is all over the siding... except for the 4 ft. x 4 ft. area I cleaned...last year!
>
> Deb Bowman

It is well known in the mold remediation industry that mold will begin to reestablish itself within twenty-four hours of using bleach. Mold will begin to grow within two weeks of applying industrial strength fungicides because they are designed to dissipate due to the fact that they pose a greater risk to human health than most of the molds they are designed to destroy. The possibility that this treatment offered long-term effects without negative side effects was intriguing.

With the science and the anecdotal evidence in hand, Ed approached his client, and to his surprise, they agreed to do the tests. We supplied the diffusers, the essential oils, and the labor. The client paid for the third-party laboratory analyses. Complete details of this first test are included in Chapter 4, Case Study No. 1.

Does It Really Matter If an Essential Oil Is Pure Therapeutic Grade?

Those who have used truly pure, therapeutic grade essential oils are horrified by the overabundance of commercials touting soaps, shampoos, perfumes, and air fresheners as aromatherapy. What these commercials are hawking is nothing more than synthetic, chemically altered, and sometimes just plain toxic aroma. There is no therapy whatsoever in that commercially hyped toxic soup. As mentioned earlier in this Chapter, to our knowledge, only a couple of companies in the United States sell what can truly be called pure, therapeutic grade essential oils. So, when it comes to essential oils: Buyer Beware!

To be absolutely clear, when we use the terms *essential oils* and *therapeutic grade essential oils* in this book, we are referring to the lipid (oil) soluble portion of the volatile, aromatic compounds obtained by steam distillation and cold expression or cold pressing of certified organic plant materials (including the stems, branches, fruits, flowers, seeds, roots, bark, needles, leaves and any other part of the plant) in a way that preserves the essential oil in a form that is as close to nature as the extraction process will permit. This definition also necessitates that the oil is a combination of compounds developed during the entire distillation process, that the oil has been distilled with loving care to avoid the high temperatures and pressures that might damage the more fragile constituents, and that every reasonable effort has been made to prevent contamination by synthetics, chemicals, preservatives and cleansers, in order to preserve the living energy and therapeutic benefits of the oil.

The essential oils found in shops and department stores, Christian bookstores, and even in most health food stores are of uncertain quality at best. At worst, they can actually be harmful. Many times essential oils are mislabeled, not necessarily through a malevolent intent to defraud but through ignorance or a rush to maximize profits. Most essential oil companies look for the lowest price with no thought to the quality of the oil, and that is why you hear about the great things essential oils can do but when you try the ones at your local retail outlet, they often do not work.

The depth of this problem was highlighted in an article published by Dr. Akash Chopra, organic chemist at the University of London, and Julian Franklin, head of Horticulture and Controlled Environment Services at Rothamsted Experimental Station, Harpenden, UK.[56]

It is a known fact that the volume of essential oils such as Rose, Lavender and Rose Geranium traded in the market is far in excess of the world-wide production of these oils. Adding synthetic chemical components, or in some cases low-value oils, makes up the volumes traded … Further adulteration occurs during the supply chain … At any stage in the supply chain, adulteration may occur when blending of essential oils or adding of specific low value chemicals can increase the value of an oil tenfold.

Here is the crux of the problem:

1. Far more essential oils are sold than are produced every year.

2. Essential oils may be adulterated at any point before they reach the consumer — from the seed to the distillery to the bottler to the retailer.

Therefore, the onus is on you, the consumer, to know that the company or person growing, distilling, bottling, and selling your essential oils is more interested in quality and benefits than in profits. Being appropriately compensated and making a profit are well deserved when you provide a high quality product that provides the benefits expected. If profits are a company's primary or singular motivation, then there is reason to search for essential oils elsewhere.

It is also important to realize that by far the largest portions of essential oils are sold as flavor enhancers and perfume components. Buyers for these industries desire oils that smell and taste exactly the same every single time. They strive to keep costs as low as possible, and they do not care whether an essential oil has any therapeutic benefit. Nature does not, however, produce exact duplicates of itself year after year. That is why wines are prized for the year in which they were produced and the vineyard in which they were grown. The components of earth, wind, sunlight, moisture, altitude, temperature, and nutrients available all contribute to making a plant's fruit, as well as its essential oil, truly unique from one year to the next. Even if you have exactly the same plant species, grown from the same seed in the same field, there will be many similarities but also minor differences in the organic compounds that make up the essential oil that plant produces from one year to the next.

The only way to ensure the same smell and the same taste year after year is to standardize an essential oil. That means adding or subtracting compounds, and doing that will alter — and may actually destroy — the therapeutic benefit of the oil.

Unfortunately, there is no current standard set forth by a governmental body called "therapeutic grade." And there are a number of essential oil companies who claim to have therapeutic grade essential oils. However, there are a few rare companies that endeavor to produce a quality of essential oil that is a grade above anything else available in the marketplace, and these essential oils provide many benefits to body, mind, and spirit. To be certain you have a truly pure therapeutic grade essential oil, ask if the manufacturer or the bottler produces them primarily for therapeutic benefit and whether they are FDA-approved as food supplements. Find out if the manufacturer and bottler meet AFNOR/ISO standards. The acronyms AFNOR (Association Francaise de Normalization) and ISO (International Standards Organization) will appear on the label of essential oils that have been independently tested and found to meet a standardized minimum profile set by the essential oils industry. The presence of these acronyms on the label does not provide certainty that the essential oils are therapeutic

quality. However, according to Bernadette Ruetsche, a spokesperson in charge of AFNOR standards for essential oils, the AFNOR/ISO standards attest to a high degree of certainty as to the purity of oils, because the AFNOR-adopted quantities of components correspond to those found in natural products. Hence, according to AFNOR authorities, if an essential oil fits AFNOR/ISO standards, it is generally considered to have come from a natural source (CARE Newsletter, Vol. 3, No. 3, April, 2005).

Ask if the manufacturer grows and distills its essential oils, and whether the oils are sent to independent laboratories for testing. In-house testing alone is not sufficient. Ask the person selling them if they use the oils themselves and whether they have received any therapeutic benefit from them. If they say they have, ask them to tell you some of the benefits they have personally experienced with the oils. Ask them if they would use the oils even if they never made a penny from them, and if they hesitate or their answer is that they are not sure, then you know not to waste your money on those essential oils.

If the oils are not pure and not therapeutic grade, not organically produced and distilled, then the oils will not provide the benefits you seek. It is that simple.

Oils that do not meet the highest standard for quality may actually do harm. In one case we are familiar with, a Canadian gentleman read in a book that Wintergreen and Cypress oils had helped eliminate knee pain. The Canadian gentleman went to his local pharmacy and found oil of Wintergreen on the shelf. After applying the oil for several days, his knee had swollen to twice the original size, and the pain had increased exponentially. The Canadian gentleman contacted the author of the book, David, telling him the Wintergreen oil had made his knee much worse. Questioning the

Canadian gentleman, David determined that the essential oil had been purchased at a pharmacy. So, David asked the gentleman to read the label and learned he had purchased synthetic oil of wintergreen, which is toxic to the body. David recommended oil from a company known to produce therapeutic grade essential oils, and after the Canadian gentleman applied the therapeutic grade essential oils of Wintergreen and Cypress to his knee, the swelling and pain were reduced in a matter of days, and the gentleman was very happy.

The above story demonstrates the importance of using high-quality essential oils when you are putting that oil on your body. If you are dealing with a mold infestation, does it really matter whether the oil is therapeutic grade or not? Yes.

First, synthetic, adulterated, or standardized oils may have no effect on mold. The tests that were completed by the authors used high-quality, therapeutic grade essential oils. Second, you will be diffusing the oil in the space you, your family, tenants, or coworkers will be occupying. If you are diffusing something that is ineffective or potentially harmful, it is at the very least a waste of your money and time, and at worst it could result in causing someone to become ill or to suffer unnecessary toxic exposure.

Always use high-quality, therapeutic grade essential oils. Do not waste your money on something that may not work or may cause harm to you, to someone you love, or to anyone else.

The Real Test

It is well known to mycologists, biologists, and professionals doing mold remediation that just killing mold is not sufficient to remediate mold. As mentioned earlier, molds produce toxins that may remain in the air even after mold has been destroyed. And dead mold spores are just as allergenic as living mold spores.

The field tests conducted by Dr. Close indicate that Essential Oil Blend No. 2 is a strong antifungal agent with powerful inhibitory effects on the most common and most hazardous toxic molds found in buildings. More than that, the tests showed that the mold spores, both living and dead, were actually removed from the air. This was one of the most surprising results of the tests and was borne out in all Case Studies completed. (For complete details, see the Case Studies in Chapter 4.)

While we are satisfied that these essential oils are very effective against molds when used in accordance with the protocol developed by Ed, we believe there is an urgent need for additional research into the use of essential oils for mold cleanup and remediation (see Chapter 5 for suggestions for additional research). We expect that it will be forthcoming. Meanwhile, we will continue to collect and publish field data and case studies in order to provide helpful information to those dealing with toxic mold who wish to address their issues in a cost-effective manner. It is anticipated that these scientifically documented case studies will be a catalyst for additional research assessing the efficacy of this treatment for buildings infested with mold.

In the coming chapters, we will separate myth from fact and poor advice from a rational approach to dealing with toxic mold. We will also discuss the need for sampling and how you can find a credentialed professional who will be able to do proper sampling and help you remediate mold before it becomes a risk to your health, your property, and your life.

References

[1] Antifungal activity of thyme (Thumus vulgaris L.) essential oil and thymol against moulds from damp dwellings, Letters in Applied Microbiology 44 (2007) 36-42, Klaric, Kosalec, Mastelic, Pieckova, and Pepeljnak.

[2] Antibacterial and Antifungal Properties of Essential Oils, published in the journal Current Medicinal Chemistry, Volume 10, Number 10, May 2003, pp. 813-829(17) Authors: Kalemba, D. and Kunicka, A.

[3] Evaluation of some essential oils for their toxicity against fungi causing deterioration of stored food commodities. Applied Environmental Microbiology. 1994 April; 60(4): 1101–1105. Authors: A K Mishra and N K Dubey.

[4] Primary cutaneous aspergillosis in human immunodeficiency virus-infected patients: Two cases and review. Clinical Infectious Disease 27:641-643, Arikan, S., O. Uzun, Y. Cetinkaya, S. Kocagoz, M. Akova, and S. Unal. 1998.

[5] Invasive aspergillosis. Clin Infect Dis. 26:781-803, Denning, D. W., 1998.

[6] Aspergillus fungemia: Report of two cases. Clin. Infect. Dis. 20:598-605, Duthie, R., and D. W. Denning. 1995.

[7] Aspergillus osteomyelitis in a child treated for acute lymphoblastic leukemia. Pediatr. Infect. Dis. J. 9:733-736, Flynn, P. M., H. L. Magill, J. J. Jenkins, T. Pearson, W. M. Crist, and W. T. Hughes. 1990.

[8] Cutaneous aspergillosis: a report of six cases. Brit J Dermatol. 139:522-526, Galimberti, R., A. Kowalczuk, I. H. Parra, M. G. Ramos, and V. Flores. 1998.

[9] Invasive Aspergillus spp infections in rheumatology patients. J Rheumatol. 26:146-149, Garrett, D. O., E. Jochimsen, and W. Jarvis. 1999.

[10] The spectrum of pulmonary aspergillosis. Journal of Thoracic Imaging. 7:56-74, Gefter, W. B. 1992.

[11] An algorithmic approach to the diagnosis and management of invasive fungal rhinosinusitis in the immuno-compromised patient. Otolaryngol Clin N Amer. 33:323-334,IX, Gillespie, M. B., and B. W. O'Malley. 2000.

[12] Invasive otitis externa due to Aspergillus species: Case report and review. Clin. Infect. Dis. 19:866-870, Gordon, G., and N. A. Giddings. 1994.

[13] Aspergillus valve endocarditis in patients without prior cardiac surgery. Medicine. 79:261-268, Gumbo, T., A. J. Taege, S. Mawhorter, M. C. McHenry, B. H. Lytle, D. M. Cosgrove, and S. M. Gordon. 2000.

[14] Combined distal and lateral subungual and white superficial onychomycosis in the toenails. J Am Acad Dermatol. 41:938-44, Gupta, A. K., and R. C. Summerbell. 1999.

[15] Ocular aspergillosis isolated in the anterior chamber. Ophthalmology. 100:1815-1818, Katz, G., K. Winchester, and S. Lam. 1993.

[16] Aspergillus meningitis in an immunocompetent adult successfully treated with itraconazole. Clin. Infect. Dis. 23:1318-1319, Mikolich, D. J., L. J. Kinsella, G. Skowron, J. Friedman, and A. M. Sugar. 1996.

[17] A case of Aspergillus myocarditis associated with septic shock. J Infection. 37:295-297, Rouby, Y., E. Combourieu, J. D. Perrier-Gros-Claude, C. Saccharin, and M. Huerre. 1998.

[18] Inhibitory action of some essential oils and phytochemicals on the growth of various moulds isolated from foods, Food Science and Technology, Evandro Leite de Souza, Evandro Leite de Souza, Edeltrudes de Oliveira Lima, Kristerson Reinaldo de Luna Freire, and Cristina Paiva de Sousa; Brazil. 2005.

[19] The Chemistry of Essential Oils Made Simple, by Dr. David Stewart, CARE Publications, 2005.

[20] The emerging role of Fusarium infections in patients with cancer. Medicine (Baltimore). 67:77-83. Anaissie, E., H. Kantarjian, J. Ro, R. Hopfer, K. Rolston, V. Fainstein, and G. Bodey. 1988.

[21] Emerging fungal pathogens. Eur. J. Clin. Microbiol. Infect. Dis. 8:323-330. Anaissie, E. J., G. P. Bodey, and M. G. Rinaldi. 1989.

[22] Microdilution susceptibility testing of amphotericin B, itraconazole, and voriconazole against clinical isolates of Aspergillus and Fusarium species. J Clin Microbiol. 37:3946-3951. Arikan, S., M. Lozano-Chiu, V. Paetznick, S. Nangia, and J. H. Rex. 1999.

[23] Atlas of Clinical Fungi, 2nd ed, vol. 1. Centraalbureau voor Schimmelcultures, Utrecht, The Netherlands. de Hoog, G. S., J. Guarro, J. Gene, and M. J. Figueras. 2000.

[24] A study of mycotic keratitis in Mumbai. Indian J Pathol Microbiol. 42:81-7, Deshpande, and Koppikar. 1999.

[25] Opportunistic fusarial infections in humans. Eur. J. Clin. Microbiol. Infect. Dis. 14:741-754. Guarro, and Gene. 1995.

[26] Spectrum of microbial keratitis in South Florida. Am J Ophthalmol. 90:38-47, Liesegang, T. J., and R. K. Forster. 1980.

[27] Fungal infections in pediatric oncology. Pediatr Med Chir. 17:435-41. Manfredini, L., A. Garaventa, E. Castagnola, C. Viscoli, C. Moroni, G. Dini, M. L. Garre, G. Manno, C. Savioli, Z. Kotitsa, and et al. 1995.

[28] Severe corneoscleral infection. A complication of beta irradiation scleral necrosis following pterygium excision. Arch Ophthalmol. 111:947-51, Moriarty, A. P., G. J. Crawford, I. L. McAllister, and I. J. Constable. 1993.

[29] Emerging pathogens. Med Mycol. 38:225-236. Ponton, J., R. Ruchel, K. V. Clemons, D. C. Coleman, R. Grillot, J. Guarro, D. Aldebert, P. Ambroise-Thomas, J. Cano, A. J. Carrillo-Munoz, J. Gene, C. Pinel, D. A. Stevens, and D. J. Sullivan. 2000.

[30] Disseminated fusarial infection in the immunocompromised host. Rev. Infect. Dis. 10:1171-1181. Richardson, S. E., R. M. Bannatyne, R. C. Summerbell, J. Milliken, R. Gold, and S. S. Weitzman. 1988.

[31] The spectrum of pulmonary infections in cancer patients. Curr Opin Oncol. 13:218-223. Rolston, K. V. I. 2001.

[32] Skin and nail infections due to Fusarium oxysporum in Tuscany, Italy. Mycoses. 41:433-437. Romano, C., C. Miracco, and E. M. Difonzo. 1998.

[33] The changing spectrum of fungal keratitis in south Florida. Ophthalmology. 101:1005-13. Rosa, R. H., Jr., D. Miller, and E. C. Alfonso. 1994.

[34] Histopathology of fungal rhinosinusitis. Otolaryngol Clin N Amer. 33:251-276,VII,VIII,NIL_5. Schell, W. A. 2000.

[35] Invasive infection with Fusarium chlamydosporum in a patient with aplastic anemia. J Clin Microbiol. 36:1772-1776. Segal, B. H., T. J. Walsh, J. M. Liu, J. D. Wilson, and K. J. Kwon-Chung. 1998.

[36] Spectrum of fungal keratitis at Wills Eye Hospital, Philadelphia, Pennsylvania. Cornea. 19:307-12. Tanure, M. A., E. J. Cohen, S. Sudesh, C. J. Rapuano, and P. R. Laibson. 2000.

[37] Disseminated zygomycosis due to Rhizopus schipperae after heatstroke. J Clin Microbiol. 37:2656-2662. Anstead, G. M., D. A. Sutton, E. H. Thompson, I. Weitzman, R. A. Otto, and S. K. Ahuja. 1999.

[38] Ten years' experience in zygomycosis at a tertiary care centre in India. J Infection. 42:261-266. Chakrabarti, A., A. Das, A. Sharma, S. Panda, S. Das, K. L. Gupta, and V. Sakhuja. 2001.

[39] Two cases of disseminated mucormycosis in patients with hematological malignancies and literature review. Eur. J. Clin. Microbiol. Infect. Dis. 17:859-863. Cuvelier, I., D. Vogelaers, R. Peleman, D. Benoit, V. Van Marck, F. Offner, K. Vandewoude, and F. Colardyn. 1998.

[40] Mucormycoses. Mycoses. 44:253-260. Eucker, J., O. Sezer, B. Graf, and K. Possinger. 2001.

[41] Brain abscess following marrow transplantation: experience at the Fred Hutchinson Cancer Research Center, 1984-1992. Clin Infect Dis. 19:402-8. Hagensee, M. E., J. E. Bauwens, B. Kjos, and R. A. Bowden. 1994.

[42] Granulomatous mediastinitis due to Rhizopus species. Am. J. Clin. Path. 70:103-107. Leong, A. S. Y. 1978.

[43] Unusual fungal pathogens in fungal rhinosinusitis. Otolaryngol Clin N Amer. 33:367-373,X. Schell, W. A. 2000.

[44] Pulmonary Rhizopus rhizopodiformis cavitary abscess in a cardiac allograft recipient. J Cardiovasc Surg (Torino). 40:223-6. Tan, H. P., A. Razzouk, S. R. Gundry, and L. Bailey. 1999.

[45] Mould Allergy, Yousef Al-Doory and Joanne F. Domson, Lea and Febiger, Philadelphia, 1984. 287 p.

[46] Manual of Medical Mycology by John Thorne Crissy, Heidi Lang, Lawrence Charles Parish, Blackwell Sciences, Cambridge, Massachusettes, 1995. 263p.

[47] Solid- and Vapor-Phase Antimicrobial Activities of Six Essential Oils: Susceptibility of Selected Foodborne Bacterial and Fungal Strains. J. Agric. Food Chem., 53 (17), 6939 -6946, 2005. 10.1021/jf050709v S0021-8561(05)00709-0. P. López, C. Sánchez, R. Batlle, and C. Nerín. Copyright © 2005 American Chemical Society.

[48] Screening for Antifungal Activity of Some Essential Oils Against Common Spoilage Fungi of Bakery Products. M. E. Guynot, S. Marĺn, L. SetÚ, V. Sanchis, A. J. Ramos
Food Science and Technology International, Vol. 11, No. 1, 25-32 (2005).

[49] Antifungal activities of essential oils and their constituents from indigenous cinnamon (Cinnamomum osmophloeum) leaves against wood decay fungi. Sheng-Yang Wang, Pin-Fun Chen, Shang-Tzen Chang, for the Dept. of Forestry, National Chung-Hsing University, School of Forestry and Resource Conservation, National Taiwan University, Taiwan, Bioresource Technology, Vol. 96, Issue 7, pgs. 813-818, May, 2006.

[50] Screening for Inhibitory Activity of Essential Oils on Selected Bacteria, Fungi, and Viruses. Chao, SC; Young, DG, Oberg, CJ. Journal of Essential Oil Research, 1998.

[51] Antibacterial properties of plant essential oils. Deans, SG; Ritchie, G. International Journal of Food Microbiology 5, 165-180, 1987.

[52] Influence of Some Spice Essential Oils on Aspergillus parasiticus Growth and Production of Aflatoxins in a Synthetic Medium. Farag, RS; Daw, ZY; Abo-Raya, SH. Journal of Food Science, Vol. 54, No. 1, 74-76, 1989.

[53] Inhibition of Growth and Aflatoxin Production in Aspergillus parasiticus by Essential Oils of Selected Plant Materials. Tantaoui-Elaraki, A; Beraoud, L. Journal of Environmental Pathology, Toxicology and Oncology; 13(1):67-72, 1994.

[54] Effect of a Diffused Essential Oil Blend on Bacterial Bioaerosols. Chao, S.C.; Young, D.G.; and Oberg. C.J. Journal of Essential Oil Research 10, 517-523 (Sept/Oct 1998).

[55] Essential Oils Desk Reference, Compiled by Essential Science Publishing, Third Edition, Third Printing, March, 2006.

[56] Integrity of Essential Oils, Positive Health Complimentary Medicine Magazine. A Chopra; J Franklin.
http://www.positivehealth.com/permit/Articles/Aromatherapy/frank55.htm.

PART THREE

THE CASE STUDIES

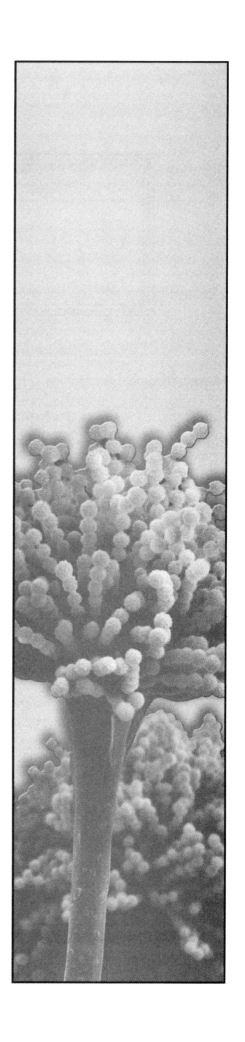

CHAPTER 4

CASE STUDIES

Introduction

It has been known for centuries that the essential oils of many plant species are antifungal, and numerous laboratory studies demonstrate this. (See the list of references in Chapter 3 for research reports on the antifungal properties of some of the individual oils in the essential-oils blends used in these case studies.) But, in order to draw valid conclusions about the effectiveness of any treatment on mold, it is good to have both laboratory *and* field studies.

Laboratory research studies are good because they are performed under controlled circumstances where the relevant physical and chemical parameters can be carefully measured and recorded, and uncontrollable variables can either be eliminated or minimized. Field studies, on the other hand, if carefully executed, yield even more valuable information because they test the effectiveness of a treatment under real-world conditions.

Mold spores are released during a certain phase in the lifecycle of molds. Therefore, the complete and accurate characterization of the level and extent of mold activity at any site requires extensive sampling and analysis performed over time. This is not always possible in the field due to client requirements, time constraints and deadlines, as well as monetary considerations. While such constraints limited the extent of data collected at many of the test sites, the sampling, analysis, and recording of procedures are sufficient to demonstrate the value and effectiveness of the blend of therapeutic grade essential oils used in these studies.

Thanks to clients who were willing to allow us to set up tests in their buildings, we are able to present what, to our knowledge, is the first comprehensive analysis of the results of field studies using therapeutic grade essential oils to treat mold infestations in buildings. We expect that additional research will be forthcoming and that it will substantiate the efficacy of this treatment.

When the first test cases were set up in a forty-eight-unit apartment complex in Southeast Missouri, in November of 2005, we identified a number of essential oils that were known to be powerful antifungal agents:

Cinnamon
Clove
Lemongrass
Melaleuca
Oregano
Thyme

When diffused for extended periods of time, most of these single essential oils are not considered appealing in smell, and all except Melaleuca may cause irritation to eyes, mucous membranes, and nasal passages.

Essential Oils and Equipment Used in These Case Studies

The therapeutic grade essential oils selected for the first three test cases were an individual or single essential oil of Thyme (*Thymus vulgaris* CT thymol), and two blends or synergies of essential oils. The blends selected were well known to be appealing in smell to most people, and both had some anecdotal evidence suggesting that they were effective antifungal agents.

Essential Oil Blend No. 1 (EOB1) is a proprietary mixture of Lemongrass (*Cymbopogon flexuosus*), Melaleuca (*Melaleuca alternifolia*, also known as Tea Tree Oil), Rosemary (*Rosmarinus officinalis* CT cineol), Citronella (*Cymbopogon nardus*), Lavandin (*Lavandula x hybrida*), and Myrtle (*Myrtus communis*).

Essential Oil Blend No. 2 (EOB2) is a proprietary mixture of Clove (*Syzygium aromaticum*), Cinnamon Bark (*Cinnamomum verum*), Lemon (*Citrus limon*), Rosemary (*Rosmarinus officinalis CT cineol*), and Eucalyptus (*Eucalyptus radiata*).

Also used in the case studies presented in this book were:

1. A 14 oz. bottle of household cleaner containing EOB2.

2. A cold-air diffuser that produced an unimpeded micro-mist of pure essential oil. This diffuser utilized a non-reactive metal well and an electric pump that forced air through the pure essential oil at

a rate of about 1 liter per minute, pushing it into a glass nebulizer. The oil molecules were atomized and dispersed into the air as a micro-mist.

3. A 1-quart spray bottle, containing the undiluted household cleaner that includes EOB2. This spray bottle was used to clean small areas of visible mold.

Photo 5: The Cold-Air Diffuser, a 15 ml bottle of EOB2 (Thieves® essential oil blend), and the Thieves® Household Cleaner.

4. A paint sprayer with a corrosion-resistant metal nozzle, utilized in spraying the undiluted household cleaner containing EOB2 in crawl spaces and over other large surfaces.

5. A high-volume vacuum pump capable of pulling from 5 to 30 liters of air per minute, used to collect spore-trap air samples. A minimum of 150 liters of air

was pumped for every air sample collected.

6. Laboratory-supplied sampling kits, including spore traps, tape-lift slides, swabs, and wall-cavity sampling kits.

7. Personal protective equipment (PPE) including: Tyvek® suit with head and shoe covering, N-95 respirator, goggles, ear plugs, and rubber gloves.

All of the therapeutic grade essential oils, the household cleaner, and the cold-air diffusers used in the case studies presented in this book were purchased from a company located in Lehi, Utah. EOB1 is sold under the name Purification®. EOB2 is sold under the name Thieves®. The household cleaner containing EOB2 is sold under the name Thieves® Household Cleaner.

Photo 6: Air-Sampling Equipment.

Extensive remodeling and construction activities were going forward at the apartment complex during the three original field tests. Crews of carpenters, electricians and plumbers continued their work even while the

diffusers were dispersing the essential oils, in order to meet tight deadlines and schedules. Workers reported that they did not like the smell of the Thyme essential oil, and they inadvertently cut off power to the diffuser in the unit where Thyme was being tested after a few hours. The power outage was not discovered until the end of the 24-hour period. Therefore, the test for Thyme oil did not provide results.

The test for EOB1 was compromised by the removal of a wall between the test unit and an adjacent unit during the diffusing stage. This flooded the test unit with dust and debris, and eliminated any possibility of meaningful comparisons between before and after samples.

This left the test of EOB2 as the only experiment with reliable data. Fortunately, as it turned out, the test data for EOB2 were very striking, and the workers reported that they very much liked the smell of EOB2. The story and the data collected during this test are presented below as Case Study No. 1.

Since the time of the first test with EOB2, largely because of the success of the test, we have had the unique opportunity to collect a considerable amount of additional data on the use and effectiveness of this blend of essential oils in connection with investigations of mold-infested buildings. More than thirty additional field studies in mold-infested buildings have been conducted. Sites investigated included private residences, warehouses, farms, rental properties, office buildings, banks, senior citizens' apartments, and health care facilities. Not all of the cases are reported in this book. Some are still on-going investigations, and others were repetitive in the information provided and were not included for that reason.

Approximately one-half of the investigations discussed in this book were prompted by specific health problems, such as headaches, respiratory problems, rhinitis, flu-like symptoms, and memory loss that were being experienced by occupants of the buildings in question. The rest were performed to meet government, bank, or other institutional inspection requirements to help property owners avoid liabilities and undue risk resulting from potential health hazards and structural damage to property.

The data presented in this book consist of third-party, certified laboratory analytical results from spore trap, tape lift, bulk, and wall probe samples. Field conditions were standardized to the extent possible. The specific sampling and analysis methods used are discussed within each case study. For an explanation and discussion of sampling methods, see Chapter 7.

In some of the case studies, multiple samples were collected immediately before and after the application of EOB2, as well as over the next several weeks or months. In other cases, due to limitations of time and/or funds, only the minimum number of samples required to document the presence or elimination of the specific mold problem of concern to the client were collected. In all cases, however, valuable data were collected demonstrating the efficiency of EOB2. As a result of these studies, we have developed a very effective protocol for treating mold with this non-toxic, eco-friendly antifungal agent. The generalized protocol is discussed in detail in Chapter 7.

Samples were collected in accordance with standard industry procedures. Times, air volumes, and other pertinent information were carefully recorded, and the samples were packed and shipped under strict chain-of-custody conditions. Samples were analyzed at Environmental Science Corp., located near Nashville, Tennessee, a U.S. Environmental Protection Agency (EPA)-certified environmental laboratory, and analyses were performed by qualified microbiologists using standard EPA methods of direct examination and viable culture techniques.

As the data files for these case studies grew, we began to think about ways to extract the maximum amount of information from the data in an easily understandable form. How much data did we have to deal with? The number of samples collected at case study sites ranged from four to sixty-five, and data derived from the samples include spore- and colony-forming unit counts for thirty-six mold species, as well as several categories of composites of mold species.

Explanation of Graphs and Tables

The data for each case study are summarized in graphs and tables within each case study discussion. Representative likenesses of actual laboratory data reports and Chain of Custody forms are presented in Appendix B. The first thirteen case studies are projects where we were able to collect complete data sets. That is to say samples were, at the very least, collected before and after treatment. Several have more. The data from these thirteen cases allow us to calculate the efficiency of treatment with the essential oil blend. All of the case studies included in this book yielded valuable data and information about mold infestation, mold-related health issues, and documentation of how the application of the essential oil blend affected those issues.

In order to see what we could learn from these data, first, we identified the mold species growing inside each case-study building from tape-lift and/or bulk sample results, and by comparing the mold species found in the indoor-air samples with mold species found in outdoor-air samples. These comparisons are represented mathematically by the ratios of indoor to outdoor mold spore concentrations.

Spore-trap air samples collected from a building that is healthy from the standpoint of mold growth will exhibit lower concentrations of each species inside the building than is found outside the building, yielding ratios that are less than one. Conversely, samples from buildings with mold infestations growing inside the building will yield ratios greater than one. In the unlikely event that the concentrations for a given species are equal inside and outside, with a ratio of 1:1, then it is possible that species is growing inside. However, additional sampling is required to confirm indoor growth, and this was not always possible in the case studies.

Indoor to outdoor ratios are calculated for the samples taken both before the application of essential oils and again after the application. Next, the spore removal efficiency (SRE) was calculated for each species of mold found growing inside the building, and for total spore counts in the building. This enabled us to determine just how effective the protocol had been for each case study. Over-all results and conclusions for all case studies are combined and summarized in the next chapter.

Spore removal efficiency (SRE) is defined simply as the percentage of available spores removed by the treatment. When there are no mold spores of a given species found in outdoor samples, the SRE is just the percentage of spores removed from the indoor air by the treatment. However, when spores of a given species are found in outdoor samples, then the exchange of air and mold spores must be accounted for in the calculation. A detailed description of the calculations appears in the section "Calculation of Spore Removal Efficiency (SRE)" at the end of this introduction to Chapter 4.

In most cases, certain factors, such as HVAC filtering efficiency and indoor/outdoor traffic through sampling areas by homeowners or construction workers during or after treatment, increase the exchange between indoor and outdoor air, affecting the number of available spores. In order to make the calculated SRE as accurate as possible, such effects are noted and taken into consideration as appropriate in each individual case study.

Initially, it was assumed that the level of exchange might vary greatly, depending upon the conditions of each site. As the case studies proceeded, the data gathered indicated that there was considerable outdoor-indoor air exchange in most of the buildings studied and that exchange also varied from species to species.

All laboratory analytical data are reported and assumed valid for purposes of case-study evaluation, unless ruled out as spurious because of events and conditions on-site that were identified at the time of treatment and/or sample collection. It is also possible for sampling and analytical errors to occur and affect ratios and other calculations, especially in cases of low spore counts. Overall, however, sampling and analysis errors probably have less effect than uncontrollable site conditions. Ambient outdoor spore counts can vary, depending on temperature, humidity, wind direction, and local activities; however seasonal variations were generally found to override other variations. All species found in each case study are shown in the tables, but only the species of greatest concern are graphed.

A description of site conditions and the timeline of collection of the first samples, treatment, and re-sampling are given with each case study. Data graphs and tables appropriate to each case study are presented, as well as observations and conclusions. While permission was granted to use the data, the actual names and exact locations for the case studies are withheld to protect the privacy and confidentiality rights of our clients.

The SRE for each species of mold found at a site is calculated and presented in the table(s) following each individual case study. The SRE could only be calculated when both before- and after-treatment samples were collected. In Chapter 5, the data from the thirteen case studies where both before- and after-treatment samples were collected are combined and the average SRE for each species of mold is calculated.

The average SRE provides a more reliable measure of the overall effectiveness of the treatment against a particular species of mold. For example, the SRE for *Stachybotrys* in Case Study No. 1 is 100%. In Case Study No. 2, however, the SRE for *Stachybotrys* is 99.9%. Looking at Table 24, on page 132 in Chapter 5, we see that in ten of the thirteen cases where *Stachybotrys* was found the SRE was 100%; however, the average SRE for this species of mold was 99.78%.

Calculation of Spore Removal Efficiency (SRE)

The SRE is equal to available spores per unit volume of air during treatment, minus spores per unit volume remaining after treatment, divided by available spores per unit volume during treatment, multiplied by 100 to convert the decimal result into a percentage.
When a given species is found in samples taken inside the space being treated but not found in outdoor samples, the available number of spores per unit volume (cubic meter of air) is given by the lab report for the indoor sample collected immediately before treatment started, and calculation of the SRE is straight forward.

Equation: $SRE = \dfrac{C_{ib} - C_{ia}}{C_{ib}} \times 100$

C = Concentration of mold spores per cubic meter of air
i = Inside
b = Before treatment
a = After treatment

If, however, the species is found in both the inside and the outdoor samples, the number of available spores per cubic meter of air during treatment is affected by the amount of exchange of air between the indoor environment and the ambient air outside.

$$SRE = \dfrac{C_{ib} + E_{avg} - C_{ia}}{C_{ib} + E_{avg}}$$

where E_{avg} = the average exchange during the treatment in mold spores per cubic meter.

This exchange can be positive or negative for a given species. It is also affected by the size, weight, and other physical features of the spores of each mold species. The number of available spores per unit volume was determined for each species and for the total spores for all species combined in each case study, to avoid to the extent possible over- or under-estimating the SREs.

CASE STUDY NO. 1 – Original Test in an Apartment Complex

This case study presents one of the three original experiments conducted in November and December, 2005, at a thirty-five-year-old, forty-eight-unit apartment complex in Southeast Missouri that had suffered leaks and flooding. The tenants had been evacuated, and the property had been closed and vacant for some time. When new owners purchased the property, they began extensive renovations and remodeling. Mold- and water-damaged ceiling tiles, insulation, and sheetrock were removed, and a professional cleaning service was hired to clean the apartments. The mold-infested apartments were first cleaned and treated with a concentrated hospital disinfectant. Close Environmental Consultants was hired by the apartment complex owners and tasked with third-party, objective determination of the effectiveness of the treatment. Lift tapes, bulk samples, and air samples identified heavy-mold and toxic-mold infestation still existing in several apartments.

After removal of additional wall and ceiling materials, cleaning with an industrial fungicide, and washing with a 3% chlorine bleach solution, re-sampling showed that viable species of *Cladosporium, Stachybotrys, Aspergillus*, and several other mold colonies still existed. While the application of concentrated hospital disinfectant, industrial-strength fungicide, and bleach had an impact on the *Cladosporium* and to a lesser extent on the *Stachybotrys, Aspergillus* actually increased in some locations during the treatments used prior to our test.

Outdoor and indoor spore-trap air samples were collected immediately prior to the test. The apartment consisted of two floors, with about 400 sq. ft. of floor space on each floor, connected by an open stairwell. A cold-air diffuser was set up in the center of the ground floor and run continuously for 24 hours, using about three-quarters of a 15 ml bottle of EOB2. Indoor and outdoor spore-trap air samples were collected again immediately after diffusing. Remodeling was ongoing at this site, with doors and windows only partially installed, and electricians and plumbers were going in and out of the test apartment, as well as spending extended periods working inside the apartment, during the 24-hour period of continuous diffusing. The workers reported that the scent of the oil was, "Okay. Kind of like cinnamon." One said he noticed his headache went away when he was working in the room with the diffuser.

> When samples were collected in the apartment at the end of two weeks, with no additional treatment, the levels of *Cladosporium* and *Aspergillus* were further reduced, and *Stachybotrys* was zero!

Graph 1 and Table 1, found at the end of this case study, depict the time line and sampling results. Note that *Stachybotrys, Penicillium/Aspergillus,* and *Cladosporium* species often referred to as black toxic mold, and the number of total spores were reduced dramatically during the 24-hour diffusing of EOB2.

The owners of the apartment complex were understandably impressed with the results, but they raised an important question, "What if the mold comes right back?"

To answer this question, the unit was left untreated for two weeks. During that time, no oils were diffused, and no further treatment was applied in the apartment. In addition, carpenters and electricians continued going in and out of the apartment during the two weeks after treatment. This would result in an exchange of air between the inside and outside, allowing the introduction of additional mold spores from the outside. When samples were collected in the apartment at the end of two weeks, the levels of *Cladosporium* and *Aspergillus* were further reduced, and *Stachybotrys* was zero!

> Indoor spore removal efficiency (SRE) was 100% for eleven species, nearly two thirds of the mold species found.

With this striking evidence of the effectiveness of the blend of essential oils, the owners authorized treatment of all apartments with any hint of mold and/or water damage. EOB2 was diffused for 24 hours or more, with the result that mold spores were reduced to acceptable levels in all apartments.

Table 1 records individual and total spore concentrations (spores per cubic meter) and indoor to outdoor ratios before diffusing and two weeks after diffusing. Indoor spore removal efficiency (SRE) was 100% for eleven species, nearly two thirds of the mold species found. Of these eleven species, nine were also present in outdoor samples. This tells us that, even with constant exchange of indoor and outdoor air, for these species, all available spores were removed. For the remaining six species, outdoor concentrations were relatively high, and the spores of these species are light and easily moved on air currents.

Considering these facts, windy conditions during the test period, unfinished doors and windows, and the workman traffic, the average available spores outside during the test period were added to the indoor concentrations to estimate the available spores

per cubic meter in the calculation of SRE for these species.

Observations and Conclusions

Laboratory results indicated that nine species were growing inside the apartment. Indoor/outdoor ratios for these nine species were greater than one.

The concentration of *Basidiospores* in the apartment was more than forty-two times the outdoor concentration. Three species, *Stachybotrys chartarum, Scopulariopsis, and Polythrincium,* were found indoors, and not found in the outside air samples, indicating their growth or presence inside the apartment.

Overall spore removal efficiency was calculated to be 98.6%. While this is impressive, it is more important to note that 100% removal was achieved for dangerous toxic mold species, including *Stachybotrys chartarum, Chaetomium, Nigrospora, and Torula.*

It is also important to note that direct-examination spore-trap analysis does not distinguish between live and dead spores. Common objections to the use of fungicides are:

1. Dead spores can cause still allergic reactions.

2. Most fungicides are toxic to humans, plants and animals.

3. The effects of fungicides wear off quickly.

All these objections are overcome with the use of this essential oil blend.

1. Both dead and live spores are removed from the air.

54

2. The essential oil blend EOB2 is nontoxic and is reported to support health and wellness. This blend is approved by the FDA for human consumption.

3. Finally, diffusing EOB2 is shown to have long-lasting residual effects.

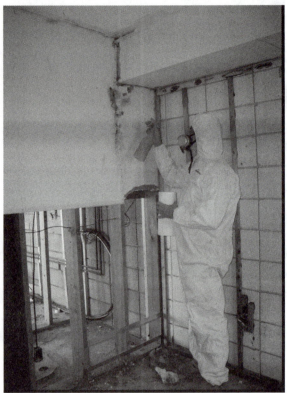

Photo 7: Personal Protective Equipment Used During Cleaning. Removal of *Stachybotrys chartarum* Colonies from Wall with EOB2 (Thieves®) Household Cleaner.

When this case was discussed with professional cleaning and mold remediation people, they were impressed by the fact that diffusing the essential oil blend for 24 hours resulted in declining spore counts for weeks after application with no further treatment. It is not unusual to see mold growth rebound within 24 hours of treatment with bleach and conventional fungicides.

The data from this case study provides clear evidence that diffusing EOB2 for as little as 24 hours efficiently eliminated toxic mold and airborne spores where bleach, hospital disinfectants, and industrial strength fungicides failed.

Overall spore removal efficiency was calculated to be 98.6%. While this is impressive, it is more important to note that 100% removal was achieved for dangerous toxic mold species, including *Stachybotrys chartarum, Chaetomium, Nigrospora, and Torula.*

GRAPH NO. 1, CASE STUDY NO. 1

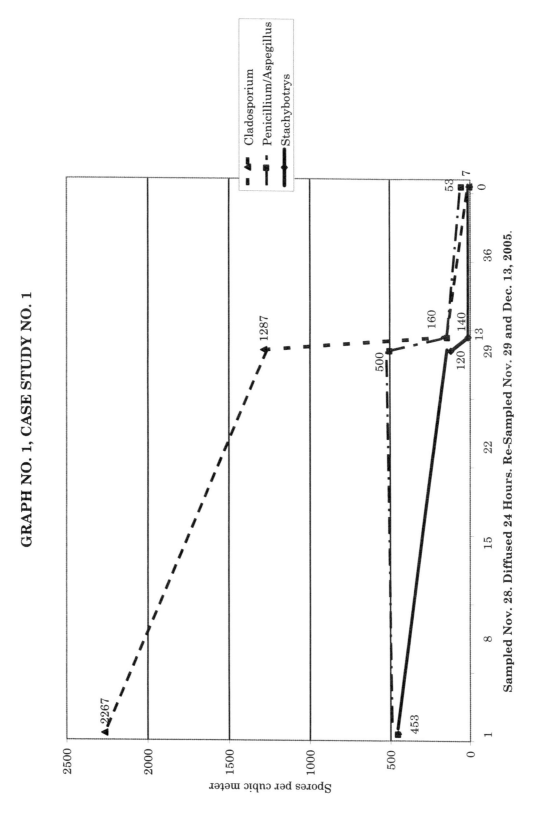

Legend:
- Cladosporium
- Penicillium/Aspegillus
- Stachybotrys

Spores per cubic meter

Sampled Nov. 28. Diffused 24 Hours. Re-Sampled Nov. 29 and Dec. 13, 2005.

TABLE 1 - CASE STUDY NO. 1

Mold Species	Cib	Cob	Cib/Cob	Cia	Coa	Cia/Coa	SRE
Alternaria	7	47	0.15	0	20	0.00	100.0%
Ascospores	1200	227	5.29	7	7	1.00	99.5%
Basidiospores	6887	163	42.25	127	233	0.55	98.2%
Bipolaris/Drechslera	27	7	3.86	0	0	NA	100.0%
Chaetomuim	80	13	6.15	0	13	0.00	100.0%
Cladosporium	1287	4303	0.30	7	173	0.04	99.8%
Nigrospora	13	13	1.00	0	13	0.00	100.0%
Other Colorless	40	40	1.00	7	63	0.11	92.3%
Other Brown	53	243	0.22	7	53	0.13	96.5%
Penicillium/Aspergillus	500	320	1.56	53	1373	0.04	96.1%
Rusts	40	20	2.00	0	0	NA	100.0%
Smuts, etc.	153	580	0.26	0	67	0.00	100.0%
Stachybotrys	120	0	GIO	0	0	NA	100.0%
Torula	7	7	1.00	0	7	0.00	100.0%
Pithomyces	13	20	0.65	0	7	0.00	100.0%
Scopulariopsis	13	0	GIO	0	0	0.00	100.0%
Polythrincium	7	0	GIO	0	0	0.00	100.0%
Total	10447	6003	1.74	208	2029	0.10	98.6%

C = Concentration (Mold Spores per cubic meter) i = inside, b = before, o = outside, a = after

SRE = Spore Removal Efficiency

Bold Indicates growth inside the building GIO = Growing inside Only

NA = Not Appropriate

CASE STUDY NO. 2 – Apartment with Highest *Stachybotrys* Levels

This case study involves another unit at the forty-eight-unit apartment complex in Southeast Missouri and was treated after the initial tests. It is presented as a separate case study because it is as different from Case Study No. 1 as any other case study carried out miles away. Sampling conducted before treatment with EOB2 and the samples taken after treatment were collected almost a month after Case Study No. 1. Also, the mix of mold species was quite different in this apartment.

The level of *Stachybotrys chartarum* spores in the before-treatment air sample from this unit was extremely high. More visible mold was found in this unit than in any other on the property. For this reason, it was decided to diffuse EOB2 continuously for 72 hours. The diffuser was set up in the center of the ground-floor portion of the apartment. This test was limited to diffusing EOB2. (See Graph 2 at the end of this case study for the time-line plots of the spore counts for the species with the highest concentrations.)

Because the conditions relating to exchange of indoor and outdoor air (i.e., unfinished doors and windows, windy weather, and construction-worker traffic) were very similar to those for Case Study No.1, the same approach was used for SRE calculations. (See Table 2 for calculation results. Indoor/outdoor ratios greater than one, indicating growth in the unit, are indicated by an asterisk.)

Observations and Conclusions

The concentrations of mold spores in this apartment were the highest found in any of the case studies. *Stachybotrys chartarum* had a spore count more than **fifteen hundred times** the outdoor concentration. Seventeen species of mold had indoor/outdoor ratios greater than one, indicating growth inside the apartment.

The *Stachybotrys chartarum* indoor-spore count dropped from 10,667 spores per cubic meter to 13 spores per cubic meter following treatment with EOB2. The total spore count in this unit was almost twelve times the outdoor levels before diffusing and was reduced to about seven-tenths the outdoor level after diffusing. Overall spore removal efficiency was 93.9%, which is slightly lower than the SRE achieved in Case Study No. 1.

While the diffuser ran three times longer than in Case Study No. 1, it only dispersed about twice as much oil, or approximately 22.5 ml of EOB2 in this unit. The reason for the lower dispersion rate is unknown. The reason for the lower SRE for the first three mold species listed in Table 2 is also not known; however, these mold species have very lightweight spores and are abundant in the outdoor ambient air. It is possible that spores were carried in by workmen at the time of the post-diffusing sampling.

> The *Stachybotrys chartarum* indoor spore count dropped from 10,667 spores per cubic meter to 13 spores per cubic meter following 72 hours diffusion of EOB2.

Chapter 4

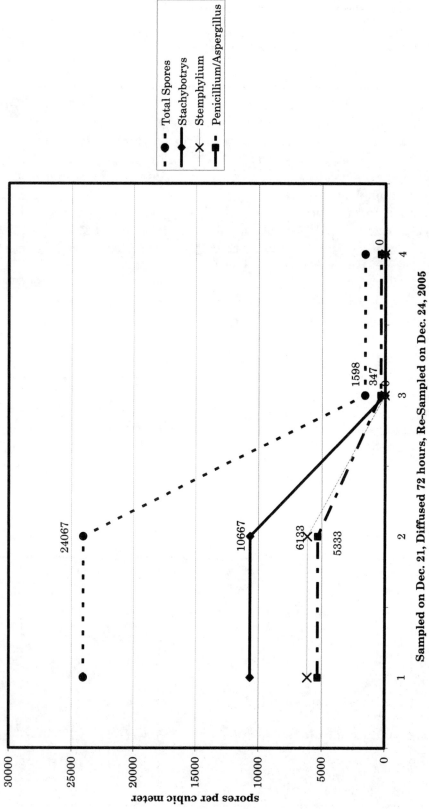

GRAPH NO. 2, CASE STUDY NO. 2
Species with above 5,000 spores per cubic meter

Legend:
- Total Spores
- Stachybotrys
- Stemphylium
- Penicillium/Aspergillus

spores per cubic meter

Sampled on Dec. 21, Diffused 72 hours, Re-Sampled on Dec. 24, 2005

TABLE 2 - CASE STUDY NO. 2

Mold Species	Cib	Cob	Cib/Cob	Cia	Coa	Cia/Coa	SRE
Alternaria	27	20	**1.35**	13	0	0.0	64.9%
Ascospores	53	7	**7.57**	267	533	0.5	17.3%
Basidiospores	320	233	**1.37**	307	740	0.41	61.9%
Chaetomuim	267	13	**20.54**	33	0	0.0	87.9%
Cladosporium	267	173	**1.54**	193	573	0.34	69.8%
Curvularia	27	0	**GIO**	0	0	NA	100.0%
Epicoccum	27	0	**GIO**	0	0	NA	100.0%
Fusarium	13	0	**GIO**	0	0	NA	100.0%
Nigrospora	27	13	**2.08**	0	0	NA	100.0%
Other Colorless	400	67	**5.97**	0	20	0.0	100.0%
Other Brown	347	53	**6.55**	33	0	0.0	91.2%
Penicillium/Aspergillus	5333	1373	**3.88**	693	340	1.02	88.8%
Smuts, etc.	133	67	**1.99**	33	7	4.74	80.6%
Spegazzinia	0	7	**0.00**	0	0	NA	100.0%
Stachybotrys	10667	7	**1524**	13	0	0.0	99.9%
Stemphylium	6133	0	**GIO**	0	0	NA	100.0%
Oidium	13	0	**GIO**	0	0	NA	100.0%
Pithomyces	13	0	**GIO**	13	0	NA	0.0%
Total	24067	2033	**11.84**	1598	2213	0.72	93.9%

C = Concentration (Mold Spores per cubic meter) i = inside, b = before, o = outside, a = after

SRE = Spore Removal Efficiency

Bold Indicates growth inside the building   GIO = Growing inside Only

NA = Not Appropriate

60

CASE STUDY NO. 3 – Apartment Providing First Residual Effects Data

This case study examines another unit in the forty-eight-unit complex, identical to the first two in floor plan and layout. This case study differs from the first two as follows: 1.) It is located in a different building, across a common area containing lawn and trees. 2.) The approach to sampling and treatment was altered based on information learned from the first two case studies. 3.) Because diffusing EOB2 had been determined to be effective at this site, the clients approved use of a household cleaner containing EOB2 for removing visible mold.

Due to the large number of units requiring treatment, and in order to save time and money, six units were treated at a time. The clients further decided that samples would not be collected before treatment beyond the first group of six, and only the most heavily infested apartment of each set would be sampled after treatment. This unit is, therefore, the only one in the complex other than the units in Case Studies 1 and 2, where before and after, inside and outdoor samples were collected.

SRE values were calculated using the same methods as Case Studies No. 1 and No. 2. The physical conditions, including openness, windy weather, and worker traffic were similar to the first two cases. As with the first two cases, the diffuser was placed in the center of the apartment on the ground floor. EOB2 was diffused continuously for 24 hours, dispersing approximately 11 ml of EOB2. Visible mold was removed from sheetrock and other finishes using the household cleaner containing EOB2.

Observations and Conclusions

Ten species of mold were found growing in this unit. Indoor/outdoor ratios greater than

one are printed in bold in Table 3. *Cladosporium* was more than twenty times the outdoor level, and *Penicillium/Aspergillus* more than thirty-eight times outdoor levels. *Stachybotrys chartarum* existed inside the unit at 500 spores per cubic meter of air, while no spores for this species of mold were found in the outdoor air sample. The total spore indoor/outdoor ratio was slightly higher than that of Case Study No. 2.

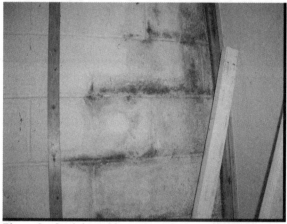

Photo 8: *Stachybotrys chartarum* and *Aspergillus* Mold Species on Concrete-Block Wall.

After-treatment samples were collected one month after diffusing was completed, yet overall SRE was 93.6%, almost identical to that of Case Study No.2 even though diffusing took place for one-third the time, and samples were taken one month after completion of treatment.

This case study provided the first indication that the residual effect of applying the essential oil blend might last for a month or longer, even where conditions conducive to mold growth remained.

Chapter 4

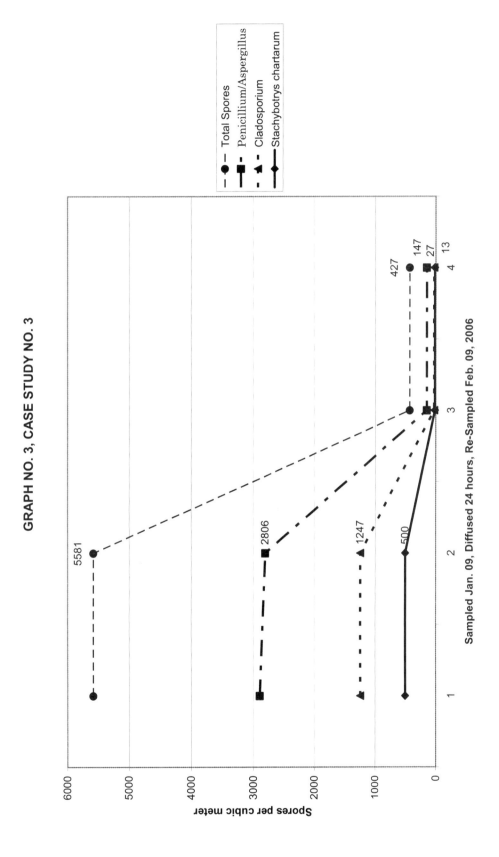

GRAPH NO. 3, CASE STUDY NO. 3

Total Spores
Penicillium/Aspergillus
Cladosporium
Stachybotrys chartarum

Spores per cubic meter

Sampled Jan. 09, Diffused 24 hours, Re-Sampled Feb. 09, 2006

TABLE 3 - CASE STUDY NO. 3

Mold Species	Cib	Cob	Cib/Cob	Cia	Coa	Cia/Coa	SRE
Alternaria	7	0	**GIO**	0	27	0.00	100.0%
Ascospores	7	7	1.00	20	240	0.08	84.7%
Basidiospores	307	47	**6.53**	93	270	0.34	80.0%
Chaetomuim	593	0	**GIO**	107	13	8.23	82.2%
Cladosporium	1247	60	**20.78**	27	267	0.10	98.1%
Nigrospora	0	0	**GIO**	0	7	0.00	100.0%
Other Colorless	27	0	**GIO**	0	53	0.00	100.0%
Other Brown	47	243	0.19	7	67	0.10	96.5%
Penicillium/Aspergillus	2806	73	**38.44**	147	867	0.17	95.5%
Smuts, etc.	33	7	**4.71**	13	7	1.86	67.5%
Stachybotrys	500	0	**GIO**	13	0	**GIO**	97.4%
Pithomyces	7	0	**GIO**	0	0	NA	100.0%
Total	5581	437	**12.77**	427	1818	0.13	93.6%

C = Concentration (Mold Spores per cubic meter) **i** = inside, **b** = before, **o** = outside, **a** = after

SRE = Spore Removal Efficiency

Bold Indicates growth inside the building   **GIO** = Growing inside Only

NA = Not Appropriate

CASE STUDY NO. 4 – Residence at an Animal Sanctuary

A three-bedroom house located on the grounds of a no-kill animal sanctuary had been vacant for several months. When the house was opened to prepare it for occupancy by a new resident caretaker, there had been leaks in the roof and around some of the windows. Carpets had been soaked in several places, and mold was visible in six locations. The closed house, temperature fluctuations, and ample available moisture resulting from leaks, humidity and condensation, provided a perfect environment for the growth of mold.

The property owner, with the help of volunteer workers, including the resident caretaker-to-be, planned to rip out the carpets and clean and paint the walls. However, because they had heard about the dangers of toxic mold, and the prospective caretaker had a history of allergic reactions to dust and mold, they decided to get professional help to determine exactly what mold species were present.

Indoor and outdoor spore-trap air samples and tape lift samples from the visible mold infestations were collected. Laboratory analytical results revealed heavy *Aspergillus/ Penicillium* growth in two of the bedrooms, *Stachybotrys chartarum* in the living room, and spores from a variety of mold species in the air. EOB2 was diffused continuously for 24 hours in the center of the 1200 sq. ft. house with all the interior doors open. About 11 ml of EOB2 were dispersed in the space. Following treatment with EOB2, water- and mold-damaged materials were removed, and

the walls were washed with the household cleaner containing EOB2. Diffusing of EOB2 commenced immediately after completion of the first sampling session; however, cleaning took place over the next several weeks. Post-treatment sampling did not take place until eleven weeks after the first sampling event.

Observations and Conclusions

Twenty-three species of mold were found at this site, more than in any other case study. Seven species had higher concentrations inside than outside, indicating that they were growing in the house, and three toxic mold species were found on tape-lift samples. All mold species found are listed in Table 4. The total spore count and the species above 5,000 spores per cubic meter of air are plotted on the first graph. The major species of concern with concentrations under 500 spores per cubic meter are plotted on the second graph. In this case study, we found that the residual effect of the protocol stretched even longer, nearly three months, and spore counts of all twenty-three species were reduced significantly.

> **Spore removal efficiency was 100% for 14 of the mold species found at this site, 11 weeks after treatment with EOB2.**

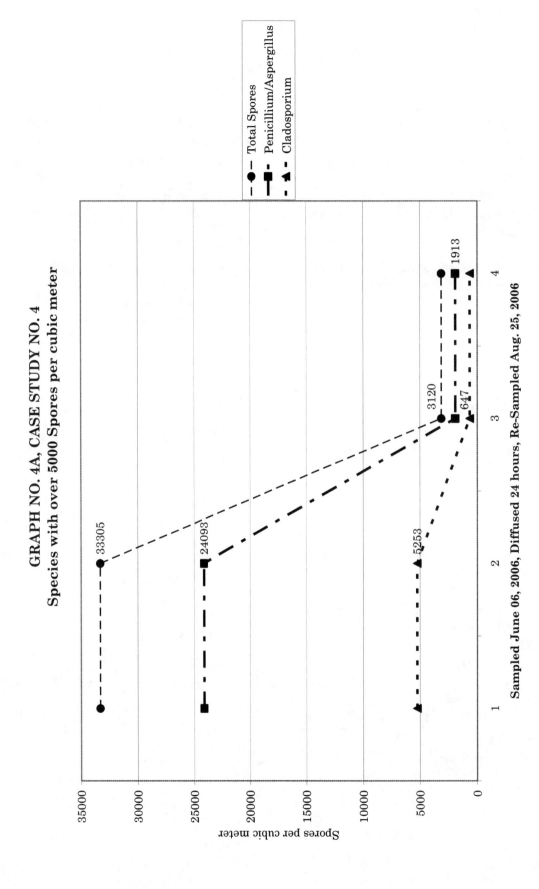

GRAPH NO. 4A, CASE STUDY NO. 4
Species with over 5000 Spores per cubic meter

Legend:
— Total Spores
— Penicillium/Aspergillus
— Cladosporium

Sampled June 06, 2006, Diffused 24 hours, Re-Sampled Aug. 25, 2006

GRAPH NO. 4B, CASE STUDY NO. 4
Species with under 500 Spores per cubic meter

Sampled 06-06-06, Diffused 24 hours, Re-Sampled 08-25-06

Spores per cubic meter

Epicoccum
Torula
Stachybotrys

TABLE 4 - CASE STUDY NO. 4

Mold Species	Cib	Cob	Cib/Cob	Cia	Coa	Cia/Coa	SRE
Alternaria	307	760	0.40	7	133	0.05	99.1%
Ascospores	847	820	**1.03**	100	1560	0.06	95.1%
Basidiospores	240	2293	0.10	373	4147	0.09	89.2%
Bipolaris/Drechslera	0	7	0.00	0	27	0.00	100.0%
Chaetomuim	53	13	**4.08**	7	0	NA	88.2%
Cladosporium	5253	7613	0.69	647	3587	0.18	94.0%
Curvularia	33	13	**2.54**	0	60	0.00	100.0%
Epicoccum	480	493	0.97	0	60	0.00	100.0%
Fusarium	27	27	1.00	40	40	1.00	33.9%
Nigrospora	0	7	0.00	0	20	0	100.0%
Other Colorless	40	287	0.14	0	113	0	100.0%
Other Brown	53	127	0.42	0	0	NA	100.0%
Penicillium/Aspergillus	24093	773	**31.17**	1913	687	**2.78**	92.3%
Smuts, etc.	1453	933	**1.56**	20	33	0.61	99.0%
Stachybotrys	27	0	**GIO**	0	0	NA	100.0%
Stemphylium	0	7	NA	0	0	NA	100.0%
Torula	240	193	**1.24**	13	13	1.00	96.2%
Cercospora	13	20	0.65	0	193	0.00	100.0%
Pithomyces	40	47	0.85	0	20	0.00	100.0%
Pestalotiopsis	13	0	**GIO**	0	0	0.00	100.0%
Scopulariopsis	0	0	NA	0	7	0.00	100.0%
Spegazzinia	93	60	**1.55**	0	0	NA	100.0%
Tetraploa	0	7	0.00	0	0	NA	100.0%
Total	33305	14500	**2.30**	3120	10700	0.29	93.2%

C = Concentration (Mold Spores per cubic meter) **i** = inside, **b** = before, **o** = outside, **a** = after

SRE = Spore Removal Efficiency   NA = Not Appropriate

Bold Indicates growth inside the building   GIO = Growing inside Only

CASE STUDY NO. 5 – Senior Citizen's Housing

Existing four-plex apartments were being converted into senior citizen housing in a small town in the Missouri Ozarks. The construction crew renovating the apartments discovered mold growing on the walls of some of the apartments. Sampling results showed that the apartments were infested with *Stachybotrys chartarum, Aspergillus, Cladosporium*, and several other mold species. These apartments were four-plex structures, with each unit having a floor plan area of about 800 sq. ft. The most heavily infested apartment was chosen for Case Study No. 5.

> **Efforts resulted in a 100% removal of 9 of the 15 mold species found growing in the four-plex, including *Stachybotrys chartarum*. No indoor/outdoor ratios were above 1 after the treatment with EOB2. Total SRE was 96.7% for this four-plex.**

Indoor and outdoor samples were collected before treatment. Water- and mold-damaged materials were removed by a contractor and EOB2 was diffused. Twenty-seven days after treatment, samples were collected again. The results are displayed in Table 5 and the graph that follows.

Observations and Conclusions

Fifteen mold species were found in the air samples collected at this site. All species had indoor/outdoor concentration ratios of one or greater, indicating that all species were growing inside the apartment, and a very high concentration of *Stachybotrys chartarum* was found inside the unit.

EOB2 was diffused continuously for 24 hours, dispersing about 11 ml of the blend of essential oils, after which visible mold was removed using the household cleaner containing EOB2. Samples were collected immediately prior to the time diffusion commenced, and twenty-seven days after cleaning efforts were completed. This resulted in 100% removal of nine of the fifteen mold species growing in the apartment, including *Stachybotrys chartarum*. No indoor/outdoor ratios were above one after the treatment. Total SRE was 96.7% for this four-plex.

68

GRAPH NO. 5, CASE STUDY NO. 5

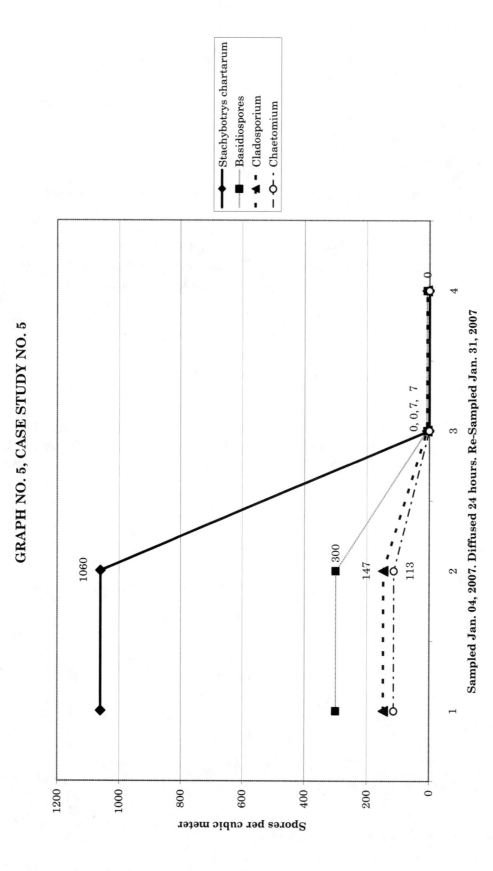

Chapter 4

TABLE 5 - CASE STUDY NO. 5

Mold Species	Cib	Cob	Cib/Cob	Cia	Coa	Cia/Cob	SRE
Alternaria	13	0	**GIO**	7	7	1.00	57.6%
Ascospores	20	0	**GIO**	0	0	NA	100.0%
Basidiospores	300	80	**3.75**	7	27	0.26	98.0%
Bipolaris/Drechslera	7	0	**GIO**	0	0	NA	100.0%
Chaetomuim	113	80	**1.41**	0	0	NA	100.0%
Cladosporium	147	93	**1.58**	7	87	0.08	97.0%
Curvularia	7	0	**GIO**	0	0	NA	100.0%
Epicoccum	40	7	**5.71**	0	0	NA	100.0%
Other Colorless	20	7	**2.86**	0	0	NA	100.0%
Other Brown	40	40	1.00	7	7	1.00	89.0%
Penicillium/Aspergillus	213	160	**1.33**	47	100	0.47	86.3%
Smuts, etc.	107	40	**2.68**	7	7	1.00	94.6%
Stachybotrys	1060	47	**22.55**	0	0	NA	100.0%
Torula	0	7	0.00	0	0	NA	100.0%
Ulocladium	7	0	NA	0	0	NA	100.0%
Pithomyces	13	0	**GIO**	0	0	NA	100.0%
Arthrinium	0	7	0.00	0	0	NA	100.0%
Gliomastix	0	7	0.00	0	0	NA	100.0%
Total	2107	575	3.66	82	235	0.35	96.7%

C = Concentration (Mold Spores per cubic meter) **i** = inside, **b** = before, **o** = outside, **a** = after

SRE = Spore Removal Efficiency   NA = Not Appropriate

Bold Indicates growth inside the building   GIO = Growing inside Only

Chapter 4

70

CASE STUDY NO. 6 – Real Estate Office

This is the case of a seventy-five-year-old Victorian style home in Texas County, Missouri, that had been converted to a real estate office. Agents working in the offices complained of headaches, coughing, and general respiratory irritation. Inspection of the building revealed numerous leaks, especially in one second-floor room and around a chimney that had been out of use since the heating had been converted from wood-burning stoves to a central air and heat system.

There was also an earth-contact basement with a sump pump. The basement was closed off from the remainder of the building, and the client did not permit entry or sampling in the basement. The earthen floor in the basement was reported to be always damp, with water in the sump.

> Spore-removal efficiency was 100% for *Stachybotrys chartarum, Chaetomium, Scopulariopsis, and Trichoderma*, with an overall SRE of 99.3%. Long-term benefits of diffusing are demonstrated by this case study.

Outdoor and indoor air samples were collected, and tape-lift samples were taken from the most prominent colonies of visible mold. This building had a number of small rooms upstairs, and several small rooms on the main floor off a larger living room and a

centrally located stairwell. The leaks were repaired, and EOB2 was diffused continuously for 24 hours in a central location on the ground floor of the building. Approximately 15 ml of EOB2 was dispersed in the space.

The agents working in the offices noticed almost immediate relief of headaches and respiratory symptoms following the diffusion of EOB2. Follow-up samples were collected seventeen days later and again five months later. There was no diffusion or mold cleaning, no removal or remediation during the interim periods. The results of sampling are displayed in Table 6 and the two graphs that follow.

Observations and Conclusions

Ten species of mold, including six known toxic species, were found growing in the building. Spore-removal efficiency was 100% for *Stachybotrys chartarum, Chaetomium, Scopulariopsis, and Trichoderma*, with an overall SRE of 99.3%. Long-term effects of diffusing on spore counts and health symptoms are also demonstrated by this case study. Even though there was a slight rebound after five months, spore counts were still far below the levels detected in the samples collected before treatment with EOB2.

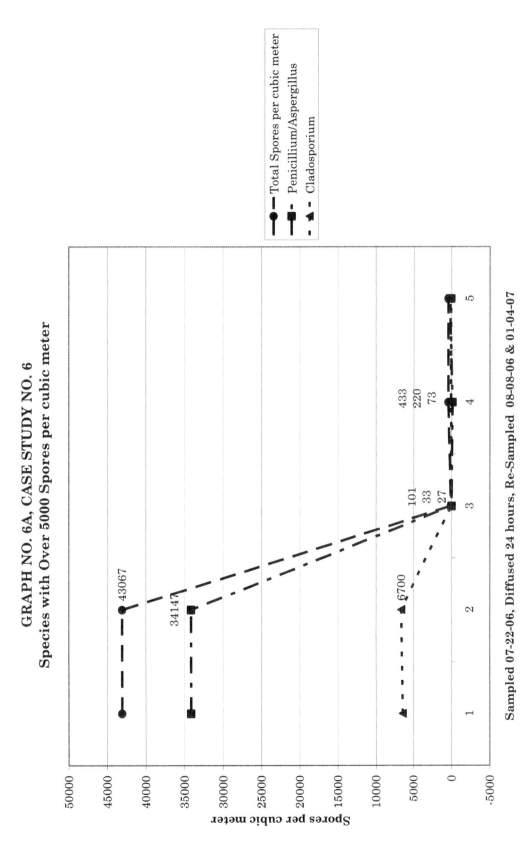

GRAPH NO. 6A, CASE STUDY NO. 6
Species with Over 5000 Spores per cubic meter

Total Spores per cubic meter
Penicillium/Aspergillus
Cladosporium

Spores per cubic meter

50000
45000
40000
35000
30000
25000
20000
15000
10000
5000
0
-5000

1   2   3   4   5

43067
34147
6700
101
33
27
433
220
73

Sampled 07-22-06, Diffused 24 hours, Re-Sampled 08-08-06 & 01-04-07

Chapter 4

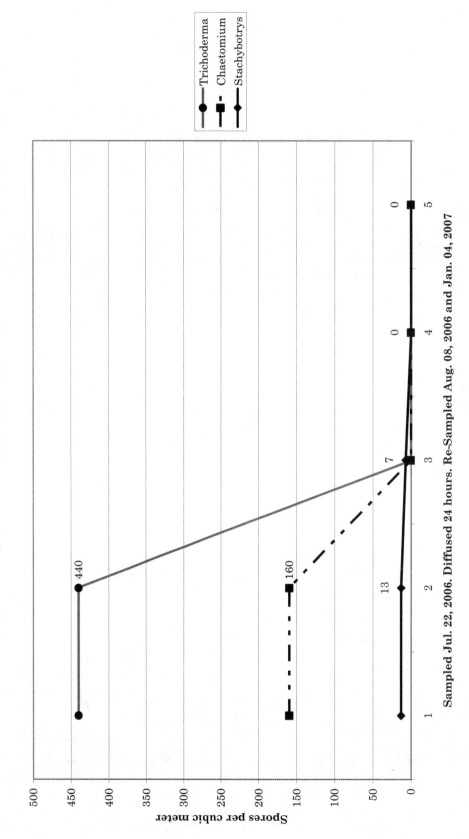

GRAPH NO. 6B, CASE STUDY NO. 6
Selected Species with under 500 Spores per cubic meter

Sampled Jul. 22, 2006. Diffused 24 hours. Re-Sampled Aug. 08, 2006 and Jan. 04, 2007

TABLE 6 - CASE STUDY NO. 6

Mold Species	Cib	Cob	Cib/Cob	Cia1	Cia2	Coa2	Cia/Cob	SRE
Alternaria	433	693	0.62	7	7	0	NA	99.1%
Ascospores	0	2940	0.00	0	13	573	0.02	99.3%
Basidiospores	267	380	0.70	7	93	20773	0.004	99.1%
Bipolaris/Drechslera	0	13	0.00	0	0	0	NA	100.0%
Chaetomuim	160	0	**GIO**	0	0	0	NA	100.0%
Cladosporium	6467	3327	1.94	27	73	633	0.12	99.1%
Curvularia	80	7	**11.43**	0	7	0	NA	91.6%
Epicoccum	220	533	0.41	0	13	27	0.48	97.4%
Fusarium	87	160	0.54	0	0	0	NA	100.0%
Other Colorless	0	7	0.00	0	0	13	0.00	100.0%
Other Brown	20	0	**GIO**	20	7	13	0.54	73.6%
Penicillium/Aspergillus	34147	440	**77.61**	33	220	240	0.92	99.4%
Rusts	0	7	0.00	0	0	0	NA	100.0%
Smuts, etc.	587	87	**6.75**	0	0	13	0.00	100.0%
Stachybotrys	13	7	**1.86**	7	0	0	NA	100.0%
Torula	13	40	0.33	0	0	0	NA	100.0%
Fusicladium	0	7	0.00	0	0	0	NA	100.0%
Scopulariopsis	13	0	**GIO**	0	0	0	NA	100.0%
Pithomyces	120	40	**3.00**	0	0	0	NA	100.0%
Spegazzinia	0	7	0.00	0	0	0	NA	100.0%
Trichoderma	440	0	**GIO**	0	0	0	NA	100.0%
Total	43067	8695	4.95	101	433	22285	0.02	99.3%

C = Concentration (Mold Spores per cubic meter) **i** = inside, **b** = before, **o** = outside, **a** = after

SRE = Spore Removal Efficiency   NA = Not Appropriate

Bold Indicates growth inside the building   **GIO** = Growing inside Only

Cib & Cob were sampled on 7-22-06; **Cia1** on 8-8-06; **Cia2** & **Coa2** on 1-4-07

CASE STUDY NO. 7 – 10,000 Sq. Ft. Office Building

A six-year-old state-agency office building, of over 10,000 square feet, developed leaks above windows in several offices shortly after it was built. A number of people working in the building complained of headaches and other symptoms typical of facilities said to have "sick building syndrome." One engineer reported that the symptoms disappeared entirely during his two-week vacation, yet returned his first day back on the job.

Agency administrators decided it was necessary to do something to protect the health of their employees. Close Environmental Consultants was hired to determine what kinds of mold species were present and to advise the client regarding remediation options.

Spore-trap, tape-lift, and bulk samples taken from four of the offices, including the office of the engineer mentioned above, and the reception area revealed *Stachybotrys chartarum* in all but one location and identified *Cladosporium, Aspergillus,* and other toxic molds in all locations. The outdoor air contained a mixture of mold species fairly typical of the area, including *Aspergillus/Penicillium.* However *Stachybotrys, Nigrospora,* and *Pyricularia* were not found outside. The presence of these species in indoor samples indicated they were growing inside the building.

EOB2 was diffused in four offices and the reception area, and two diffusers were placed inside the intake of the air-handling systems on the roof of the building. Approximately fourteen 15 ml bottles of EOB2 were diffused during the 72-hour period. Table 7A and Graph 7A represent the mold growth present on bulk-source samples obtained from the engineer's office before and after diffusing EOB2 continuously for 72 hours and following cleaning of visible mold with the

household cleaner containing EOB2. Cleaning was done by agency employees, and water was added to the cleaner against our recommendations. The data presented in the tables and charts for this case study are all related to the engineer's office.

Bulk samples (samples of material) were collected from sheetrock exhibiting visible mold growth. An approximately two-inch square of sheet rock was removed from the area on the wall that was heavily infested with visible mold and located near the engineer's desk. The bulk sample was sealed in plastic and shipped to a qualified laboratory for culture and analysis. The lab analysis showed 75,000 mold-colony-forming units of *Stachybotrys chartarum.* After diffusing EOB2 in the room and cleaning the wall, a second square of sheetrock was cut from the wall adjacent to the location of the first sample, and sent to the same laboratory for analysis. The result was zero live toxic mold colonies, indicating the treatment had been very effective.

Because the mold species involved were toxic and at such high levels, we recommended diffusing a second time for at least 48 hours before cleaning other mold-infested areas with the household cleaner and removing water-damaged materials. Workers were advised to use respirators, goggles, rubber or nitrile gloves, and protective clothing while performing cleaning and removal activities.

As a result of using this treatment, less than half of the sheetrock originally planned for removal actually had to be removed. Both the agency occupying the offices and the building contractor were pleased with the results. The agency is continuing to diffuse the essential oil blend until all leaks and sources of excess moisture are found and corrected.

Before and after air sampling results are presented in Graph B, and the calculations of indoor/outdoor ratios and SRE are presented in Table 7B.

Observations and Conclusions

In contrast with previous case studies, some of the data collected during this case study appeared puzzling. The lab report on the tape lift sample collected in this office noted "Very heavy *Stachybotrys chartarum* mold growth," and the viable bulk sample showed 75,000 *Stachybotrys* colony-forming units. However, the air samples collected in this office contained no *Stachybotrys* spores, even though *Stachybotrys chartarum* spores were found in air samples from two of five other locations sampled.

Stachybotrys chartarum spores are relatively heavy, sticky spores, so they are not as easily airborne as lighter spores, such as *Aspergillus* and *Cladosporium*. And spores for any species of mold are only released during certain phases of the mold's life cycle. Therefore, one explanation for why *Stachybotrys* spores were not found in the air sample in the engineer's office may be that the *Stachybotrys* fruiting structures were not sporulating at the time the spore-trap air sample was collected.

Another interesting feature of the analytical results in this case study was that the lab results showed five mold categories (species or groups) that were definitely growing inside the office. They were: *Stachybotrys, Nigrospora,* smuts, "other brown," and *Pyricularia*. These fungi are often associated with agriculture and, in the case of *Pyricularia,* with rice farming in particular. Rice is a major crop in the Southeast Missouri Bootheel region,

where the office building is located, and finding these species in air samples was not unexpected.

> **Twenty-three species and groups were found in the air samples collected in this office. The SREs for seventeen of these were 100%, and the overall SRE for all species and groups was 98.1%.**

In addition, the viable bulk samples collected from the engineer's office indicated that *Penicillium* levels increased following treatment. Possible causes for this occurrence include sampling error, sample contamination, lab error, and/or the presence of water used by workers who cleaned the visible mold. While a single sample does not constitute conclusive evidence, this may explain why the SRE is lower for *Penicillium/Aspergillus* (71.9%) in this case study, since the bulk sample results indicated that four-tenths of the *Penicillium/Aspergillus* spores may have been *Penicillium*.

In spite of these unique features that raise interesting questions, the analytical results and calculations show that the protocol of diffusing, cleaning, and removal of water-damaged materials was very effective. Twenty-three species and groups were found in the air samples collected in this office. The SREs for seventeen of these were 100%, and the overall SRE was 98.1%.

The engineer reported a noticeable reduction in headache and allergy symptoms immediately following the diffusion of EOB2.

GRAPH NO. 7A, CASE STUDY NO. 7, VIABLE BULK SAMPLE

Legend:
- Stachybotrys
- Aspergillus versicolor
- Penicillium
- Cladosporium

Colony-Forming Units

Sampled Jan. 10, Diffused 72 hours, Cleaned & Re-SAmpled Jan. 20, 2006

TABLE NO. 7A - CASE STUDY NO. 7, BULK SAMPLES

Mold Species	CFU Before	CFU After	% Mold CFUs Killed
Aspergillus ochraceus	2,000	0	100%
Aspergillus versicolor	4,000	0	100%
Basidiomycetes	15,000	0	100%
Cladosporium	3,000	0	100%
Non-sporulating fungi	1,000	0	100%
Penicillium	4000	3300	17.50%
Stachybotrys chartarum	75,000	0	100%
Yeasts	300	0	100%
TOTAL	104,300	3300	96.80%

Notes:

CFU = Viable Colony Forming Units

Sheetrock samples removed from the same area before (Jan.10, 06) and after (Jan. 20, 06) application of the Protocol

Chapter 4

78

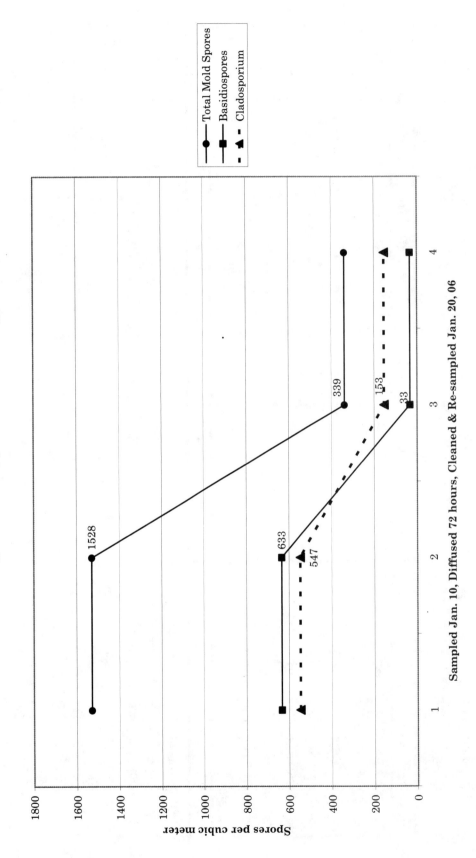

GRAPH NO. 7B, CASE STUDY NO. 7,

Sampled Jan. 10, Diffused 72 hours, Cleaned & Re-sampled Jan. 20, 06

TABLE 7B - CASE STUDY NO. 7

Mold Species	Cib	Cob	Cib/Cob	Cia	Coa	Cia/Cob	SRE
Alternaria	0	133	0.00	0	143	0.00	100.0%
Ascospores	133	1600	0.08	0	1730	0.00	100.0%
Basidiospores	633	8000	0.08	33	8640	0.00	99.6%
Bipolaris/Drechslera	0	7	0.00	0	13	0.00	100.0%
Cladosporium	547	5333	0.10	153	5760	0.03	97.5%
Curvularia	7	13	0.54	0	20	0.00	100.0%
Epicoccum	7	47	0.15	0	53	0.00	100.0%
Fusarium	0	20	0.00	0	27	0.00	100.0%
Nigrospora	7	0	**GOI**	0	7	0.00	100.0%
Other Colorless	0	240	0.00	20	320	0.06	92.9%
Other Brown	20	0	**GOI**	0	0	NA	100.0%
Penicillium/Aspergillus	93	320	0.29	120	347	0.35	71.9%
Rusts	7	20	0.35	0	27	0.00	100.0%
Smuts, etc.	47	13	**3.62**	13	20	0.65	79.5%
Cercospora	20	67	0.30	0	7	0.00	100.0%
Gliomastix	0	7	0.00	0	73	0.00	100.0%
Pyricularia	7	0	**GOI**	0	7	0.00	100.0%
Fusicladium	0	13	0.00	0	7	0.00	100.0%
Oidium	0	7	0.00	0	13	0.00	100.0%
Peronospora	0	7	0.00	0	13	0.00	100.0%
Pithomyces	0	13	0.00	0	7	0.00	100.0%
Scopulariopsis	0	7	0.00	0	0	NA	100.0%
Total	1528	15867	0.096	339	17234	0.02	98.1%

C = Concentration (Mold Spores per cubic meter) i = inside, b = before, o = outside, a = after

SRE = Spore Removal Efficiency   NA = Not Appropriate

Bold Indicates growth inside the building   GIO = Growing inside Only

80

CASE STUDY NO. 8 – A Hospital and NICU

This case study took place in a hospital. Essential oils have been used for years in hospitals and health-care facilities in Europe, especially in France, where the use of therapeutic-grade essential oils is much more of a mainstream medical practice. In the United States, essential oils were abandoned for other types of agents in the late 1800s and early 1900s, and only in the last thirty years have scientists begun to rediscover the health benefits of essential oils. For more information on hospitals using essential oils, see Appendix C.

Hospital administrators requested that samples be taken in several areas, including operating rooms, intensive care units, x-ray rooms, break rooms, and offices, specifically those frequented by a nurse who was having allergic reactions to something in the hospital. She would break out in a rash fifteen or twenty minutes after arriving at work. Her doctor was not able to determine the cause and suggested that it might be a reaction to mold. Extensive sampling found mold spores in these locations but not of sufficient concentrations or species types to cause such reactions. The cause was finally determined to be her extreme sensitivity to toxic chemicals in a solvent that was used to clean the operating rooms.

The story would have ended there, except for the fact that a few *Stachybotrys* spores were found in an air sample from a doctor's station located adjacent to a neo-natal intensive-care unit (NICU). This concerned the hospital's doctors and administrators for several reasons: First, *Stachybotrys* spores are suspected of being dangerous to persons with compromised or undeveloped immune systems, especially newborn babies. Second, *Stachybotrys* spores are rarely, if ever, found in air samples unless there is growth nearby,

because *Stachybotrys* spores are heavy, sticky, and not easily airborne. Third, this was a clean-room area, in which air pressure and oxygen content were controlled, and anyone entering had to wear a breathing mask and disposable protective clothing, and wash his/her hands with disinfecting soap upon entering the area. Furthermore, the air-handling system was protected with HEPA[+] filters that should have stopped mold spores from entering the building through the air exchange of the HVAC system.

After some detective work, including taking air samples from the overhead space above the ceiling tiles and from the HVAC return and measuring the air flow into and out of the NICU, we found that the spores were coming into the area from the overhead space, due to an ineffective seal put in place during remodeling of the NICU room. The spores were then traced to a stairwell. A leak had developed where the flat roof abutted with the stair well, allowing water to run down the inside of the stairwell wall. When the drywall was removed from the stairwell wall, large areas of black mold were exposed.

The water-damaged and mold-infested drywall was removed, and EOB2 was diffused continuously for 24 hours in the stairwell and in the NICU. The results are displayed on the graph and table that follow.

Observations and Conclusions

This was an interesting case study from the standpoint that while solving one problem – that is, finding the cause of the nurse's allergic reaction – another problem, potentially a much more serious one, was uncovered. The analytical data represented in the graph and table are for the area just outside the door to

Chapter 4

the NICU where the problem was traced to the overhead space and stairwell.

> **The overall spore removal efficiency of the essential oils protocol utilized at this hospital was 99.3%.**

Some intermediate samples were taken, after diffusing in the NICU, but before the stairwell and overhead were treated and resealed. Those results showed a high level of exchange between the indoor and outside air. The after sample used in the graph and table is representative of the conditions after the project was completed. While mold spores were still found in every room tested, the levels were extremely low (10 to 60 total mold spores per cubic meter), and there were no *Stachybotrys* spores.

The indoor/outdoor ratios in the before sample are almost all less than one for all but three species, and they had very low counts. Under most circumstances, this would not indicate a cause for concern over potential mold exposure. However, under these circumstances, in a hospital where patients typically have weak immune systems, finding even one *Stachybotrys* spore is cause for concern. We would be negligent if we did not point out that the same is true anywhere there are infants, elderly, or people recovering from surgery, antibiotics, or other drugs.

The overall efficiency of the essential oils protocol again proved to be very good at 99.3% removal. The three spore types with SRE values less than 100% are light, easily air-borne spores that are abundant outside and are often carried inside on clothes and hair.

GRAPH NO. 8, CASE STUDY NO. 8

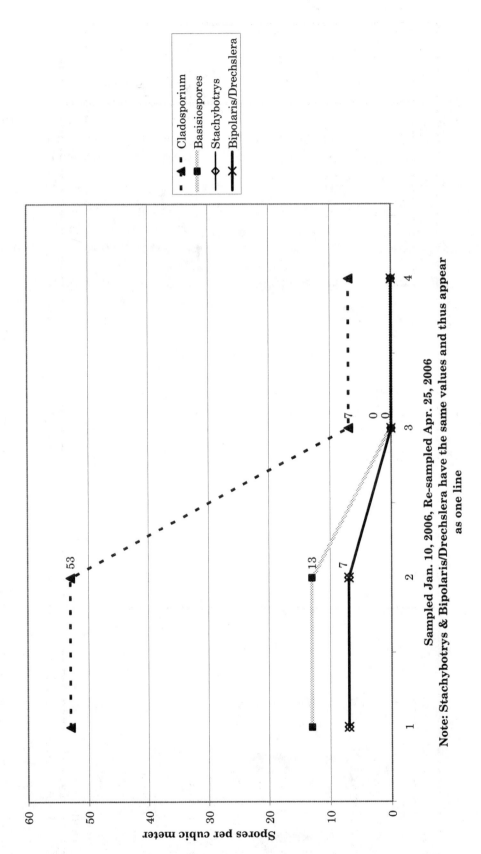

Sampled Jan. 10, 2006, Re-sampled Apr. 25, 2006
Note: Stachybotrys & Bipolaris/Drechslera have the same values and thus appear as one line

Legend:
- Cladosporium
- Basisiospores
- Stachybotrys
- Bipolaris/Drechslera

Spores per cubic meter

TABLE 8 - Case Study NO. 8

Mold Species	Cib	Cob	Cib/Cob	Cia	Coa	Cia/Coa	SRE
Ascospores	0	187	0.00	0	800	0.00	100.0%
Basidiospores	13	2987	0.004	0	3467	0.00	100.0%
Bipolaris/Drechslera	7	0	**GIO**	0	27	0.00	100.0%
Cladosporium	53	2267	0.023	7	4533	0.00	99.8%
Epicoccum	0	13	0.00	0	60	0.00	100.0%
Other Colorless	0	80	0.00	0	47	0.00	100.0%
Other Brown	13	0	**GIO**	7	27	0.26	73.6%
Penicillium/Aspergillus	13	267	0.049	40	1233	0.03	94.8%
Smuts, etc.	13	187	0.07	0	7	0.00	100.0%
Stachybotrys	7	0	**GIO**	0	0	NA	100.0%
Torula	0	13	0.00	0	20	0.00	100.0%
Cercospora	0	13	0.00	0	20	0.00	100.0%
Spegazzinia	0	7	0.00	0	0	NA	100.0%
Trichoderma	0	93	0.00	0	0	NA	100.0%
Total	119	6114	0.019	54	10241	0.01	99.3%

C = Concentration (Mold Spores per cubic meter) i = inside, b = before, o = outside, a = after

SRE = Spore Removal Efficiency

GIO = Growing inside Only

Bold Indicates growth inside the building

NA = Not Appropriate

CASE STUDY NO. 9 – Farm Home Built Over an 1830's Structure

This is the case of a farm home in an area just west of the Mississippi River in Perry County, Missouri. The home was built on top of the native stone foundation and basement of a much older structure, the original homestead, built circa 1830. The basement had a small area of earth contact that was originally used for coal storage.

People living in the home, especially the wife, a registered nurse, started having strange health problems that including constant aching-all-over, frequent flu-like symptoms, fatigue, and alarming confusion and mental decline. Her doctor prescribed a number of drugs over a period of time, but nothing seemed to work. After coming in contact with black mold in the basement, the lady of the house became very ill and had to be rushed to the hospital emergency room with a strange group of symptoms that included seizures. Ultimately these were deemed to be allergic reactions.

We were asked to investigate and found *Stachybotrys*, along with several other species of mold growing in the house. See Table 9.

Two diffusers were set up, one in a central location on the main floor of the home and one in the earth contact area of the basement. Both diffusers were run continuously for 24 hours diffusing about 22 ml of EOB2 in the space. After treatment, the woman reported that her symptoms noticeably improved within the first day. Diffusing was followed by cleaning visible mold with the household cleaner containing EOB2, and the cleaner was also sprayed into wall cavities during remodeling activities.

Observations and Conclusions

The SREs were 100% for seven species of mold found in the home, however, *Chaetomium* and *Epicoccum* SREs were not as good at this location as had been seen in other case studies. The reason for this is not known with any certainty; however, it may be related to the house having been built over an earth-contact basement with rough, unsealed rock walls. In spite of this, **the overall SRE for all species found at the site was 96.1%.**

It was recommended that the owners continue to diffuse EOB2 on a regular basis, using timers to diffuse the blend for fifteen minutes every four hours. Based on previous experience, this practice will prevent mold species from recurring.

A few weeks after this case study was completed, the woman contacted us again to say how happy she was with the results of this treatment, that she was thrilled to finally have her life back. Many of her debilitating symptoms had disappeared and she had hope for the first time in years that she would regain her health entirely.

The overall SRE for all species found at this site was 96.1%.

GRAPH NO. 9, CASE STUDY NO. 9

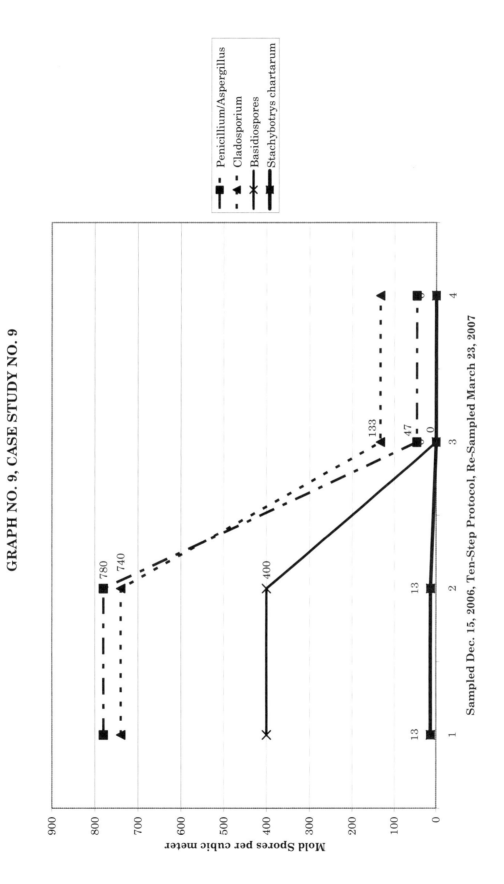

TABLE 9 - CASE STUDY NO. 9

Mold Species	Cib	Cob	Cib/Cob	Cia	Coa	Cia/Coa	SRE
Alternaria	27	13	**2.08**	7	7	1.00	81.1%
Ascospores	53	73	0.73	20	187	0.11	89.1%
Basidiospores	400	3620	0.11	0	27	0.00	100.0%
Chaetomuim	7	0	**GIO**	20	40	0.50	25.9%
Cladosporium	740	1660	0.45	133	1160	0.11	93.8%
Epicoccum	13	7	**1.86**	7	33	0.21	78.8%
Other Colorless	0	13	0.00	0	7	0.00	100.0%
Other Brown	20	27	0.74	0	20	0.00	100.0%
Penicillium/Aspergillus	780	827	0.94	47	63	0.75	96.2%
Rusts	7	0	**GIO**	0	7	NA	100.0%
Smuts, etc.	53	60	0.88	0	13	0.00	100.0%
Stachybotrys	13	0	**GIO**	0	0	NA	100.0%
Torula	7	0	**GIO**	0	0	NA	100.0%
Total	2120	6300	0.34	234	1564	0.15	96.1%

C = Concentration (Mold Spores per cubic meter) i = inside, b = before, o = outside, a = after

SRE = Spore Removal Efficiency   GIO = Growing inside Only

Bold Indicates growth inside the building

NA = Not Appropriate

CASE STUDY NO. 10 – Private Bootheel-Area Residence

This house had a minor roof leak around a rock fireplace. It also had a basement, which is unusual in this area due to a high water table that ranges seasonally from four to eight feet below land surface. The basement was equipped with a sump pump to handle seepage, and the sump always had standing water in it, providing a never-ending source of excess moisture. This house also had two earth-contact crawl spaces.

While renovating the house, workers found mold growing on the ceiling around the fireplace, under carpets on the main floor, and heavy growth on the outside of HVAC ducts in the basement. The owner contacted us to identify the mold species and advise her regarding cleanup and removal of the mold.

Her husband and two family pets had died of cancer during the past year, and renovations at the home were being undertaken to allow her daughter's family to move in to the home. She was an outdoor type of person, who spent little time inside the house, and had not noticed any adverse health-related symptoms that she attributed to mold. Her husband and pets had spent almost all of their time in the house during the past year, due to illness, and the woman wondered if the mold might have contributed to their untimely deaths.

Although it is not possible to know whether the cancers were caused by mold, there is sufficient evidence in the scientific record that links some mold species to diseases, including cancer, and death. More information on this subject is provided in both Chapters 2 and 3.

Observations and Conclusions

Air and tape-lift samples showed at least eight species of mold growing in the house, including *Stachybotrys chartarum, Aspergillus, Chaetomium, Epicoccum,* and high levels of *Graphium* were found growing on and under the carpet. All of these species are toxic, or potentially toxic, and have been linked by scientific research to serious diseases including cancers.

Because the total floor space of the house was more than 2000 sq. ft., EOB2 was diffused continuously in a central location on the main floor and also in the basement for 48 hours, dispersing four 15 ml bottles of the blend. Visible mold and the crawl spaces were sprayed with the cleaner containing EOB2. The results are displayed in the table and graphs that follow. Graph 10A depicts *Penicillium/Aspergillus* and total mold spores. Graph 10B depicts other toxic mold species present in the home.

There were some anomalies in this data: A low level of *Scopulariopsis* was found after treatment, even though none had been found in the samples collected prior to treatment. Also, *Cladosporium*, other brown spores, and smuts were higher in the post-treatment samples than had been experienced at other sites. Ongoing renovation activities and a sump pump malfunction that had flooded the basement floor on the day post-treatment samples were collected may have contributed to these unusual results.

The treatment used at this location included diffusing EOB2 continuously for 48 hours and spraying visible mold and crawl spaces with the household cleaner containing EOB2. The treatment proved very effective. The SREs for seven species and groups of molds were 100%, and the overall SRE was 99.1%.

GRAPH NO. 10A, CASE STUDY NO. 10
Species with Over 10000 Spores per cubic meter

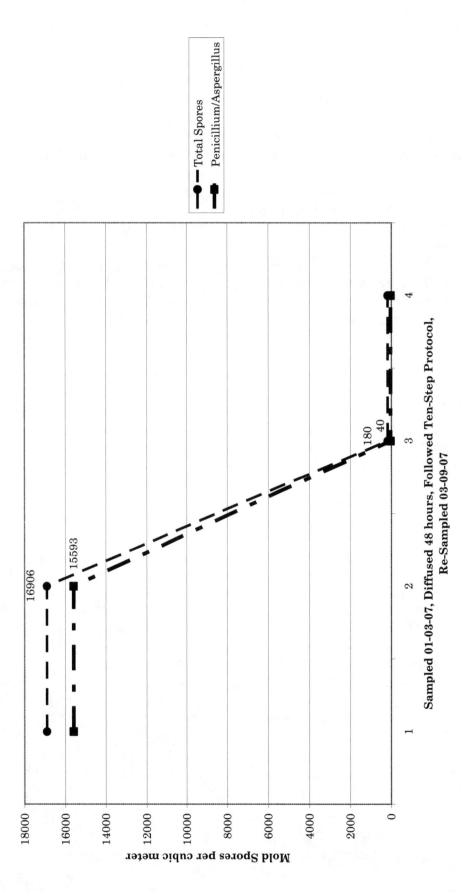

GRAPH NO. 10B, CASE STUDY NO. 10
Species with under 300 Spores per cubic meter

Sampled Jan. 03, 2007. Diffused 48 Hours and Sprayed. Re-Sampled Mar. 09, 2007

Mold Spores per cubic meter

Chaetomium
Torula
Epicoccum
Stachybotrys chartarum

293
133
60
13
0
0
0

TABLE 10 - CASE STUDY NO. 10

Mold Species	Cib	Cob	Cib/Cob	Cia	Coa	Cia/Coa	SRE
Alternaria	27	0	**GIO**	0	13	0.00	100.0%
Ascospores	180	387	0.47	0	47	0.00	100.0%
Basidiospores	87	2553	0.034	33	847	0.039	98.2%
Chaetomuim	293	0	**GIO**	0	7	0.00	100.0%
Cladosporium	420	433	0.97	80	120	0.667	88.5%
Epicoccum	60	0	**GIO**	0	7	0.00	100.0%
Other Colorless	53	33	**1.61**	0	0	NA	100.0%
Other Brown	27	0	**GIO**	20	7	**2.86**	34.4%
Penicillium/Aspergillus	15593	73	**213.60**	40	107	0.374	99.7%
Smuts, etc.	20	0	**GIO**	7	13	0.538	73.6%
Stachybotrys	13	0	**GIO**	0	0	NA	100.0%
Torula	133	0	**GIO**	0	0	NA	100.0%
Total	16906	3479	**4.86**	180	1168	0.15	99.1%

C = Concentration (Mold Spores per cubic meter) i = inside, b = before, o = outside, a = after

SRE = Spore Removal Efficiency

Bold Indicates growth inside the building   GIO = Growing inside Only

NA = Not Appropriate

CASE STUDY NO. 11 – An Immaculate Private Residence

Upon entering this beautiful, spacious home, it would be hard for anyone to imagine that there could be a mold problem. It was immaculate, with no visible signs of mold anywhere. The couple who lived in the home was still in the prime of their lives. The wife was a business owner who had retired early, and the husband had suffered a debilitating work-related back injury. Consequently, both spent considerable time in the house.

They had had an addition built onto their home and had the HVAC system reconfigured during the renovations. Not long after that project had been completed, both husband and wife began suffering from chronic headaches, dizziness, nausea, and chronic sinus problems which they attributed to stress and weather conditions.

On a routine maintenance visit, their HVAC repairman found black mold growing on a panel covering the A-coil in the central air conditioning unit. We were called in to determine what kind of mold it was.

> **In spite of the fact that there was no obvious visible mold, this case study had higher mold-spore concentrations for two toxic mold species and higher total mold spores than any other case study, even those where an abundance of visible black mold was found. This underlines the importance of sampling.**

A bulk sample taken from the HVAC panel proved to be a very heavy growth of *Cladosporium*, and a spore-trap air sample collected near outlet vents on the main floor,

taken with the HVAC system running, revealed at least five species of mold growing inside the house, including *Stachybotrys, Curvularia,* and *Penicillium/Aspergillus.* Concentrations of *Penicillium/Aspergillus* and *Cladosporium* spores were very high at 38,773 and 19,573 spores per cubic meter, respectively, with total mold spores more than three hundred times the total in the outdoor sample.

> **Laboratory results showed dramatic reductions of the extremely high spore counts found at this site. Samples taken more than one month after treatment with EOB2 showed indoor spore levels remained far below outdoor levels.**

The mold-infested HVAC panel was sprayed with the household cleaner containing EOB2 and one additional 15 ml bottle of EOB2 was added to it.

Two diffusers on the main floor dispersed EOB2 continuously over a period of more than 72 hours, using approximately 60 ml of the essential oil blend. Post-treatment samples were collected six days later. The post-treatment indoor air sample was collected in the same location, with the HVAC system operating, as had been done when collecting the original indoor sample.

Observations and Conclusions

Laboratory results showed dramatic reductions of the high spore counts to near zero. See Table 11 for indoor/outdoor ratios

92

and SRE values, and the time-line graphs that follow. Another set of post-treatment samples taken about one and one-half months later showed the indoor spore levels remained far below outdoor levels.

The occupants of the home reported that their symptoms were markedly reduced and that they very much enjoyed the smell of the essential oil blend. The husband commented that he liked to inhale the oil directly from the diffuser because it seemed to clear his head.

In spite of the fact that no visible mold was apparent at this site, this case study had higher mold-spore concentrations for two toxic mold species and higher total mold spores than any other case study undertaken, including those studies where an abundance of visible black mold was found on walls, ceilings, baseboards, and floors. This underlines the importance of sampling.

If you are suffering from allergies and chronic symptoms for which your doctor cannot identify a cause or treat effectively, and you think it might be related to your home, workplace, or other building that you frequent, air sampling may identify the source of your problem.

The couple in this case noticed an improvement in their symptoms within 24 hours of the time the treatment commenced. They remained inside the home throughout the treatment period, and they have decided

to continue diffusing the oil blend at least once a week as a preventative measure to protect their home and their property.

The results of diffusing and spraying in this case study were excellent. Twelve mold species and groups were found in the before-treatment air samples. The SREs for eight of these were 100%, with an overall SRE of 99.8%.

> **The results of diffusing and spraying in this case study were excellent. Twelve mold species and groups were found in the before-treatment air samples. The SREs for eight of these were 100%, with an overall SRE for total spores of 99.8%.**

Chapter 4

GRAPH NO. 11A, CASE STUDY NO. 11
Species with Over 19,000 Spores per cubic meter

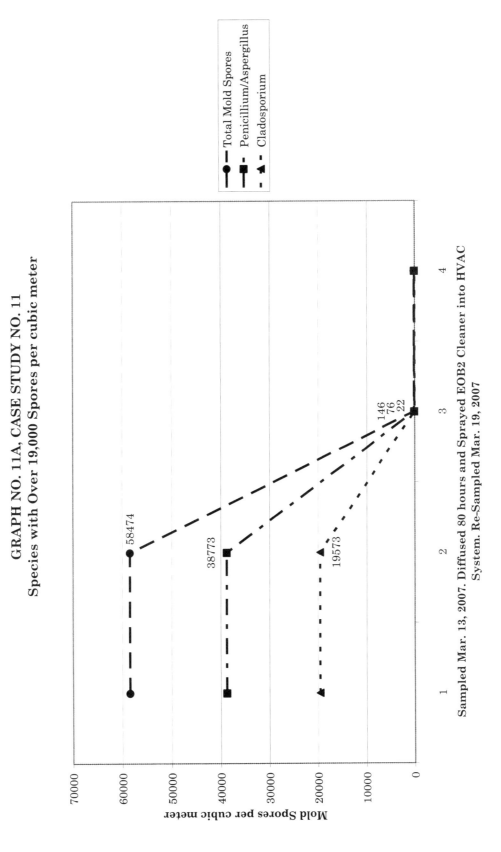

Sampled Mar. 13, 2007. Diffused 80 hours and Sprayed EOB2 Cleaner into HVAC
System. Re-Sampled Mar. 19, 2007

Total Mold Spores
Penicillium/Aspergillus
Cladosporium

Mold Spores per cubic meter

GRAPH NO. 11B, CASE STUDY NO. 11
Species with Under 50 Spores per cubic meter

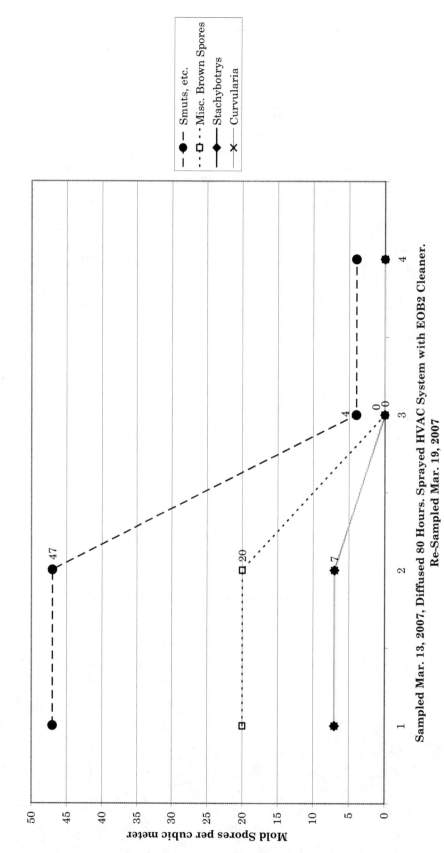

Sampled Mar. 13, 2007, Diffused 80 Hours. Sprayed HVAC System with EOB2 Cleaner.
Re-Sampled Mar. 19, 2007

TABLE 11 - CASE STUDY NO. 11

Mold Species	Cib	Cob	Cib/Cob	Cia	Coa	Cia/Coa	SRE
Ascospores	0	7	0.00	0	87	0.00	100.0%
Basidiospores	40	107	0.37	44	2747	0.02	97.0%
Cladosporium	19573	140	**139.81**	22	713	0.03	99.9%
Curvularia	7	0	**GIO**	0	0	NA	100.0%
Epicoccum	7	13	0.54	0	13	0.00	100.0%
Fusarium	0	0	NA	0	7	0.00	100.0%
Other Colorless	0	93	0.00	0	93	0.00	100.0%
Other Brown	20	147	0.14	0	147	0.00	100.0%
Penicillium/Aspergillus	38773	273	**142.03**	76	140	0.54	99.8%
Smuts, etc.	47	33	**1.42**	4	27	0.15	94.8%
Stachybotrys	7	0	**GIO**	0	0	NA	100.0%
Torula	0	20	0.00	0	20	0.00	100.0%
Total	58474	833	**70.20**	146	3994	0.04	99.8%

C = Concentration (Mold Spores per cubic meter) i = inside, b = before, o = outside, a = after

SRE = Spore Removal Efficiency

Bold Indicates growth inside the building   GIO = Growing inside Only

NA = Not Appropriate

CASE STUDY NO. 12 – Single-Family Rental Property

When the renters moved out of this single-family rental residence, the mother complained that her small children had endured constant colds and flu-like symptoms, and that she suffered from chronic headaches while in the house. Because the house was only a few years old and appeared to be in good condition, the owner thought the young mother had an overactive imagination. However, before preparing it for a new renter, he decided to have the property checked for mold.

Upon inspection, the house appeared to be very clean and in good repair. There was a little discoloration on the silicone seals in the bathroom, and only a slight hint of mold in one corner. Buckling floors were found in the hallway between the garage and the kitchen, and this area also had a slight musty odor. The buckling floor was adjacent to the laundry room, and when the panel covering the washer-dryer plumbing hook-up was removed, black mold was found covering the back of the panel. Apparently, one or both of the washer connections had been leaking.

Tape-lift and air samples showed heavy infestations of *Penicillium Aspergillus*, *Stachybotrys chartarum* and *Chaetomium* growing in the house. The *Penicillium/Aspergillus* spore count was more than **eighty-two** times the outdoor concentration, and *Stachybotrys* and *Chaetomium* levels were also very high. See Table 12 and the graph that follows. EOB2 was diffused continuously for 24 hours and the cleaner containing EOB2 was applied to mold-infested materials in the laundry room and hallway. Following initial treatment, water-damaged materials were replaced, followed by another 24 hours diffusion of EOB2. A total of 30 ml of EOB2 was dispersed in the house.

Observations and Conclusions

The spores of the three toxic mold species, *Stachybotrys, Penicillium/Aspergillus*, and *Chaetomium* totaled over 36,000 spores per cubic meter in this case study. It is likely that anyone breathing this mixture for an extended period of time would suffer adverse health effects.

Sampling results, displayed in the table and on the graph, show that the essential oil protocol was very effective, with an SRE of 98.6%.

The owner of this rental property was quite happy with the results achieved by use of the essential oils protocol developed for this site. With our report, he had documentation of the treatment and the results, showing that the mold problem had been addressed and eliminated.

> **Sampling results show that the essential oil protocol was very effective, with an SRE of 98.6%.**

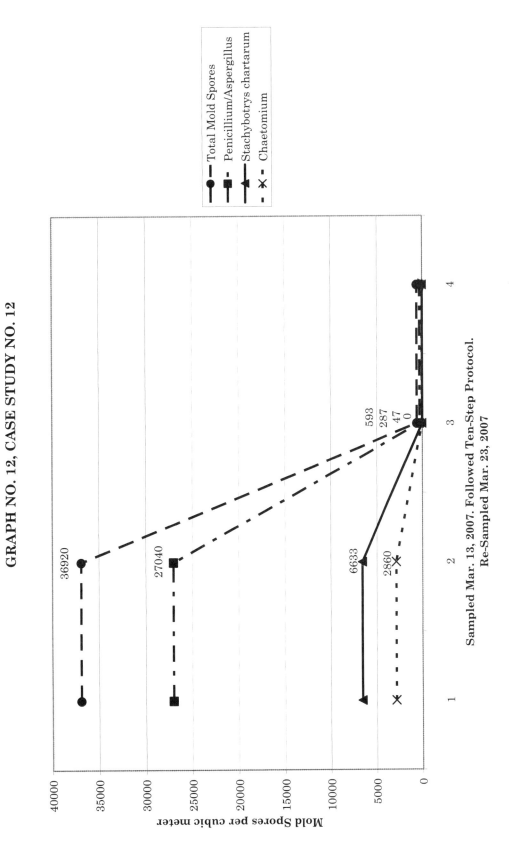

GRAPH NO. 12, CASE STUDY NO. 12

Mold Spores per cubic meter

Sampled Mar. 13, 2007. Followed Ten-Step Protocol.
Re-Sampled Mar. 23, 2007

Total Mold Spores
Penicillium/Aspergillus
Stachybotrys chartarum
Chaetomium

36920
27040
6633
2860
593
287
47
0

TABLE 12 - CASE STUDY NO. 12

Mold Species	Cib	Cob	Cib/Cob	Cia	Coa	Cia/Coa	SRE
Alternaria	0	20	0.00	0	7	0.00	100.0%
Ascospores	87	33	**2.64**	0	1547	0.00	100.0%
Basidiospores	40	273	0.15	73	4987	0.01	97.3%
Chaetomium	2860	0	**GIO**	47	0	**GIO**	98.4%
Cladosporium	240	820	0.29	60	1093	0.05	95.0%
Epicoccum	20	7	**2.86**	0	7	0.00	100.0%
Other Colorless	0	0	NA	113	0	NA	NA
Other Brown	0	0	NA	13	0	NA	NA
Penicillium/Aspergillus	27040	327	**82.69**	287	187	1.53	98.9%
Smuts, etc.	0	3673	0.00	0	93	0.00	100.0%
Stachybotrys	6633	0	**GIO**	0	0	NA	100.0%
Ulocladium	0	20	0.00	0	0	NA	100.0%
Pithomyces	0	7	0.00	0	0	NA	100.0%
Tetraploa	0	7	0.00	0	0	NA	100.0%
Scopulariopsis	0	0	NA	0	153	0.00	100.0%
Oidium	0	7	0.00	0	0	NA	100.0%
Total	36920	5194	7.11	593	8074	0.07	98.6%

C = Concentration (Mold Spores per cubic meter) i = inside, b = before, o = outside, a = after

SRE = Spore Removal Efficiency   **GIO** = Growing inside Only

Bold Indicates growth inside the building

NA = Not Appropriate

CASE STUDY NO. 13 – Private Residence with Asthmatic Cat

After purchasing a house in Southeast Missouri, this client found mold growing under wallpaper in her basement when she started a remodeling project. Since moving into the house, she had experienced more frequent headaches and allergies, and one of her cats had developed asthma.

Spore-trap air samples were collected in two basement rooms. Very high *Aspergillus/Penicillium* counts were found in the room where the wallpaper had been partially removed. *Stachybotrys chartarum* spores were found in another room where the carpet had been soaked from a leaking water heater.

Table 13 displays the species found in two indoor air samples and their concentrations. Outdoor samples were not taken at the client's request in order to limit expense. The SRE percentages were calculated as if there were no air exchange between inside and outside, because it is not possible to estimate the exchange without outdoor samples. This explains the negative SRE value for *Basidiospores*, one of the most abundant outdoor species in this area.

Even one *Stachybotrys chartarum* spore in an air sample is reason for concern because *Stachybotrys* spores are "sticky" spores that are not easily airborne. And *Stachybotrys chartarum* has been linked to serious health problems. The homeowner told us the cat, which had developed asthma since moving into the home, slept most of the day in the room where *Stachybotrys* was detected. The very high concentration of *Aspergillus/Penicillium* spores in the other room was also reason for concern, since some *Aspergillus* species, including *Aspergillus versicolor* and *Aspergillus flavus,* produce toxins that are linked to disease, and these mold species are known to

cause *Aspergillosis* (infections caused by *Aspergillus* spores) in immune-suppressed people and animals. (See Chapters 2 and 3 for more information about studies that identify health-related impacts of mold exposure.)

We referred the client to a reputable mold-remediation contractor for removal of toxic mold-infested materials from her home. Diffusion of EOB2 commenced immediately after the before-treatment samples were collected.

Air-extraction equipment was brought in by the remediation contractor to induce a negative air pressure and also an air scrubber was set up to prevent the spread of mold spores and other particulate matter while the remediation contractor removed the *Stachybotrys chartarum*-infested carpet from the room that had been flooded by the water-heater failure. The remediation contractor further recommended removing the sheetrock from the back of the closet where the *Stachybotrys* growth had been found and removing all mold-infested wallpaper in the other room using similar procedures.

When the client received the remediator's estimate for this, she called us distraught because of the expense she had already incurred and the unsettling aspect of substantial additional expenses required to complete the remediation.

We proposed additional sampling for several reasons:

1. Sampling in the first room was needed to confirm complete removal of *Stachybotrys chartarum.*

2. Not all species of *Aspergillus* and *Penicillium* are toxic. Different sampling procedures would provide the information required to determine what the client's next steps should be.

Because there was reason to suspect that there might be mold growth behind the closet wall in the room where the *Stachybotrys* colonies had been found, we proposed drilling small holes into the wall, approximately the size of a pencil's width, between the studs to pull air samples from inside the wall cavity. This would provide the information needed to determine whether the wall actually had to be removed. Bulk samples of the mold-infested wallpaper in the second room would be collected, and results from these samples would tell us whether or not the mold species growing there were toxic.

Observations and Conclusions

The graphs below display the results of spore-trap samples taken in the room with the water-damaged carpet before and after removal of the carpet and diffusing EOB2 for 24 hours.

Of the sixteen mold species and groups found in the air samples, nine were removed entirely, with SREs equal to 100%. Overall SRE for total spores was 93%.

Fortunately, no mold growth was found inside the wall cavity in the first room, and no toxic species of *Aspergillus* were found in the second room. This was extremely good news for the homeowner. By approving the extra sampling, she spent about $500 and saved several thousand dollars. In addition, she reported to us that within a week of diffusing EOB2, her cat's asthma symptoms disappeared.

> **By approving extra sampling, this homeowner spent about $500 and saved several thousand dollars. In addition, she reported to us that within a week of diffusing EOB2, her cat's asthma symptoms disappeared.**

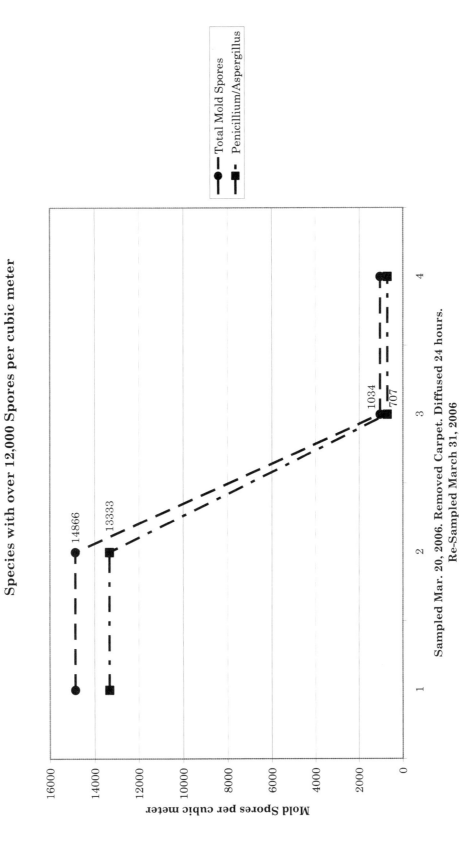

GRAPH NO. 13A CASE STUDY No.13
Species with over 12,000 Spores per cubic meter

Sampled Mar. 20, 2006. Removed Carpet. Diffused 24 hours.
Re-Sampled March 31, 2006

Mold Spores per cubic meter

Total Mold Spores
Penicillium/Aspergillus

14866
13333
1034
707

GRAPH NO. 13B, CASE STUDY No. 13
Species with 800 spores per cubic meter or less

Sampled Mar. 20, 2006. Removed Carpet. Diffused 24 hours.
Re-Sampled Mar. 31, 2006

- - - ◀ Cladosporium
- - □ - Chaetomium
—— ◆ Stachybotrys chartarum

TABLE 13 - CASE STUDY NO. 13

Mold Species	Room A	Room Bb	Room Ba	SRE
Alternaria	7	20	0	100%
Ascospores	0	53	7	86.70%
Basidiospores	0	53	60	-13%
Bipolaris/Drechslera	0	7	0	100%
Chaetomuim	0	240	67	72.10%
Cladosporium	533	800	127	84.10%
Curvularia	7	0	0	100%
Epicoccum	13	20	0	100%
Nigrospora	0	7	0	100%
Other	7	133	53	60.20%
Penicillium/Aspergillus	26,667	13,333	707	94.70%
Smuts, etc.	133	160	13	91.90%
Stachybotrys	0	13	0	100%
Stemphylium	0	7	0	100%
Pestalotiopsis	7	7	0	100%
Pithomyces	0	13	0	100%
Total	27374	14866	1034	93.00%

Notes:

Room A - mold behind wallpaper

Room B- water-damaged carpet

Bb -Before Treatment, **Ba** = After Treatment

Bold: Species suspected to be growing inside

CASE STUDY NO. 14 – Bar and Grill

This is the case of a popular bar and grill located in downtown St. Louis, Missouri. The manager, who happened to be several months pregnant, experienced headaches, chronic congestion and flu-like symptoms while at work. When she went away on a one-week vacation, she was fine, but all her symptoms returned by the end of her first day back at work. She noticed black mold in a storage room, and we were contacted to investigate.

Visual inspection identified black mold growing on drywall and on a carpet that was damp from the drain of an ice-making machine. Indoor and outdoor samples were collected. Results showed that *Stachybotrys chartarum, Chaetomium, Penicillium/Aspergillus, Cladosporium,* and several other species of mold were growing inside the building. Analytical results are presented in Table 14.

Diffusers were set up immediately following collection of the before-treatment samples. The client was familiar with the essential oil blend and had her own supply of EOB2 and a diffuser on-hand.

Because she was pregnant, we recommended that EOB2 be diffused continuously for 24 to 72 hours, and that she not be present in the room with the diffuser for more than 15 minutes at a time. We also recommended that she contract a mold-remediation company to spray and clean all visible mold with the undiluted household cleaner containing EOB2 and to remove all water- and mold-damaged materials. We also recommended that she have a plumber re-route the ice machine drainage and that she take other necessary measures to eliminate any excess moisture in the room, such as emptying drip pans and cleaning filters on the ice-machine at least once a week.

Observations and Conclusions

The manager reported that after just 24 hours of diffusing EOB2 in the storage room, her headaches, nausea, and congestion had disappeared. A professional mold-remediation company had been contracted to remove damaged drywall and clean all mold-infested areas. Follow-up samples were not collected at the request of the client. Having reviewed the data and graphs we had collected from three previous case studies and being familiar with the essential oils we were using, she saw no need to spend the money for post-treatment samples. She informed us that she was very happy and thankful for our help, and that she would diffuse again for 24 to 72 hours following all clean-up and removal activities, and would continue diffusing in the storage room on a regular basis.

TABLE 14 - CASE STUDY NO. 14

Mold Species	Ci	Co	Ci/Co
Alternaria	7	7	1.00
Ascospores	67	53	**1.26**
Basidiospores	133	200	0.67
Bipolaris/Drechslera	0	7	0.00
Chaetomuim	13	0	**GIO**
Cladosporium	467	200	**2.34**
Epicoccum	7	7	1.00
Nigrospora	7	0	**GIO**
Other Colorless	13	13	1
Other Brown	13	7	**1.86**
Penicillium/Aspergillus	433	53	**8.17**
Smuts, etc.	0	7	0.00
Stachybotrys	53	0	**GIO**
Scopulariopsis	7	13	0.54
Total	1220	567	**2.15**

Notes:

Ci = Spores per cubic meter inside

Co = Spores per cubic meter outside

Bold: Species growing inside

GIO = Growing Inside Only

CASE STUDY NO. 15 – Physical Fitness Center

A physical fitness center experienced a flood of sewage and rainwater in the basement when a sewer line was breached during construction and street-improvement activities along the edge of the property. Unfortunately, the breach occurred late on a Friday afternoon, and the stop-gap measures put in place by the construction crews were overwhelmed during several periods of heavy rainfall that occurred over the weekend, causing prolonged flooding of the building's basement.

After extensive and expensive remediation efforts undertaken by a mold-remediation contractor, the owner asked us, as an objective third party, to assess the status of cleanup. Table 15 displays the wide range of mold species, including *Stachybotrys, Chaetomium, and Penicillium/Aspergillus*, that were found growing inside the building even after all the contractor's efforts.

We recommended diffusing EOB2 and cleaning in accordance with the protocol we had developed, but due to litigation over the sewer-line break, the client decided to suspend all mold-remediation activity.

> **This case study offers compelling evidence that the EOB2 protocol is far superior to, and far more cost-effective than, conventional mold-remediation methods. The EOB2 protocol actually works. Conventional methods do not.**

Observations and Conclusions

This case study is included in order to compare the results obtained using conventional remediation techniques with the results obtained when using EOB2 in accordance with the protocols we have developed during the tests presented in this book. (A generalized 10-Step Protocol is presented and explained in detail in Chapter 7.)

Conventional techniques employed at this location included negative-air-pressure equipment, air scrubbing for several days before, during and after removal of all water-damaged carpets, drywall, and other damaged, stained or mold-infested materials. Mold-infested areas were also cleaned and treated with toxic, chemical fungicides, followed by several more days of air scrubbing. After all this effort and expense, *nine* mold species and groups, including *Stachybotrys chartarum, Chaetomium*, and other toxic mold species, were still found growing inside the building.

Contrast these conventional mold-remediation results with the results obtained during case studies presented in this book where heavily infested buildings were treated with EOB2 and after-treatment sample results indicated that **NO** species were growing inside the building (see Case Studies No. 1, 6, 7, 9, and 11).

TABLE 15 - CASE STUDY NO. 15

Mold Species	C i	Co	Ci/Co
Alternaria	53	160	0.33
Ascospores	167	533	0.31
Basidiospores	940	4213	0.22
Bipolaris/Drechslera	33	13	**2.54**
Botrys	7	7	1.00
Chaetomuim	200	0	**GIO**
Cladosporium	1160	4000	0.29
Curvularia	53	47	**1.13**
Epicoccum	27	67	1.00
Fusarium	7	133	0.05
Nigrospora	33	53	0.62
Other Colorless	80	240	1.00
Other Brown	33	33	1.00
Penicillium/Aspergillus	693	400	**1.73**
Rusts	13	27	0.48
Smuts, etc.	173	207	0.84
Stachybotrys	67	0	**GIO**
Torula	27	13	**2.08**
Zygomycetes	13	0	**GIO**
Cercospora	133	107	**1.24**
Gliomastix	7	7	1.00
Pithomyces	33	20	**1.65**
Pestalotiopsis	7	20	0.35
peronospora	0	13	0.00
Arthrinium	0	20	0.00
Trichothecium	7	7	1.00
Fusariella	0	7	0.00
Total	3966	10347	**0.38**

Notes:

Ci = Spores per cubic meter inside

Co = Spores per cubic meter outside

Bold: Species growing inside   **GIO** = Growing Inside Only

CASE STUDY NO. 16 – Private Residence, High Sensitivity

This case study is included because it illustrates the complexity of health and personal relationship issues that may be caused or aggravated by mold exposure. It also is somewhat typical of cases in which one member of a family has health problems that may be related to mold exposure while other members do not.

A couple who attended our toxic mold seminar in February of 2006 approached us after the presentation: At retirement age, the wife was having increasingly vexing health problems. She was unable to stand to do housework for more than a few minutes at a time, and she was unable to drive farther than a couple of miles without experiencing "complete exhaustion." The husband knew something was wrong, but having no significant health problems himself, he thought a lot of it might be "in her head."

When one person in a home is more susceptible to or highly allergic to mold because of a suppressed immune system, and another person living in the same space experiences no ill effects or difficulties with mold whatsoever, the situation can become a source of mistrust and contention — even ridicule.

The woman's frequent visits to doctors and specialists yielded vague diagnoses, such as fibromyalgia and chronic fatigue syndrome. She had lost the hearing in one ear, had frequent bladder infections, digestive problems, sleeping problems, and chronic flu-like symptoms. She also had deep lacerations on all of her fingertips that appeared as if they might have been deep paper cuts that never healed, but she reported that she had not had any paper cuts. A long list of drugs prescribed by her doctors brought only temporary relief, if any, and her symptoms continued to get worse year after year. Long-term and possibly overuse of antibiotics may have suppressed her immune system and contributed to her being more highly susceptible to mold.

Their home was in a rural area, at the edge of a very small town, and was extraordinarily clean. **Visual inspection did not find mold anywhere.** At the client's request, we only sampled inside, to see if any toxic mold species were inside their home. Results are displayed in Table 16. Since no outdoor samples were collected, no ratios or SREs could be calculated for this case study.

Relatively low concentrations of mold spores were found in their living room and bedrooms, with only slightly higher levels in the bathroom and basement. There were no *Stachybotrys* spores, and levels of *Penicillium/Aspergillus* were not high enough that they would ordinarily pose a major concern, although individuals who have developed sensitivities to mold may be affected by even limited spore counts, and this may have been the case in this instance.

Certain species of concern, including *Chaetomium*, *Cladosporium*, and *Curvularia*, were found in the basement. Her symptoms did suggest the possibility that she could be highly sensitive to these mold species, but there was no conclusive evidence that this was the case. Due to the storage of bulbs and other plant material and cardboard boxes in the basement, smuts, *Myxomycetes* and *Periconia* were somewhat elevated.

She had exhausted many years with doctors of various types, and blood tests a doctor had ordered, which she showed us, indicated that mold spores were present in her blood stream. That was her reason for coming to the seminar.

We recommended diffusing EOB2 for 24 hours in the basement and on the ground floor, followed by regular cleaning in the kitchen, bathrooms, and laundry areas with the household cleaner containing EOB2. It was also suggested that they continue diffusing EOB2, as a preventative measure, at least fifteen to forty-five minutes a day, or once a week for eight hours in each room of the house. Additionally, Jacqui suggested that the woman consider incorporating an aromatherapy and food supplement program to provide assistance to her body in regaining strength and energy and to support the healthy function of her immune system. The woman decided to give this a try for a month or so, just to see if she received any noticeable benefit.

Observations and Conclusions

As we have said before, everyone is different. Individuals may react differently to mold, as in this case, and may also react differently to essential oils.

In Lauri's Story, found in Chapter 2 of this book, Lauri told us that she would sit in a closed office with the diffuser on her desk, filled with EOB2 and running continuously for eight hours every day. She said doing this made her feel better. In Case Study No. 11, the gentleman living in the immaculate home reported that he felt better when he got right next to the diffuser and inhaled the essential oil blend for as long as possible. He would even sleep with a diffuser running right next to him because he felt better, clearer, and better able to function when he did that.

On the other hand, the woman in this case study reported that she was extremely sensitive to the essential oil blend. She said

that during the first treatment, she had to leave her home and stay at a local motel for the 24-hours the diffusers were running. She also reported that she was able to return to her home within one hour after the diffusers were turned off. And, surprisingly, she reported that she has not experienced any difficulty being in her home while diffusers are running for short periods of time, since that first day of intensive diffusing.

Follow-up sampling, conducted after the 24 hour treatment, showed that mold species had been reduced in the living quarters, and the lady of the house also reported significant improvements in her health. She was able to stand longer without fatigue. She had been able to drive about 75 miles, one-way, visit for several hours with a friend, and drive back home without fatigue. She told us she had not been able to do that for years.

Later, after about two months of diffusing regularly and using the supplements, the woman called to report that her doctor and therapist had reduced, by almost half, the number of prescription medications she had to take on a regular basis. She said her doctors had also reduced the frequency of her office visits and follow-ups. She told us she was very impressed and surprised by the results she had received using the essential oils and supplements, because years of medications had not provided this level of improvement in her quality of life.

This case study is important because it offers the hope that these particular essential oils, when used with common sense and attention to the needs of the individuals involved, may provide a long-term mold-prevention option that is suitable for even the most sensitive and immuno-depressed individuals.

TABLE 16 - CASE STUDY NO. 16

Mold Species	Living Area	Bathroom	Bedroom A	Bedroom B	Basement
Alternaria	7	0	0	7	7
Ascospores	0	67	7	13	13
Basidiospores	100	107	147	80	13
Chaetomuim	0	0	0	0	**13**
Cladosporium	**73**	**140**	**80**	**193**	**47**
Curvularia	**7**	0	**7**	0	**13**
Epicoccum	**7**	**7**	**0**	**7**	**7**
Nigrospora	0	0	0	0	**7**
Other Colorless	0	0	13	7	0
Other Brown	20	13	53	20	13
Penicillium/Aspergillus	120	167	87	67	207
Rusts	73	0	0	0	0
pithomyces	7	0	0	0	**13**
Smuts, etc.	53	7	13	33	**113**
Tetraploa	0	0	0	0	7
Total	467	508	407	327	473

BOLD = Species suspected to be growing inside

CASE STUDY NO. 17 – Assisted-Living Facility

Noticing water damage and suspicious discoloration in hallways, common areas, and some living units, the chief maintenance engineer of an assisted-living facility contacted us to take samples and determine whether the facility had a mold problem. Twelve indoor samples were collected from common areas, dining rooms, an office, kitchens, hallways, and living quarters, along with one outdoor air sample. The indoor samples included eight spore-trap air samples and four tape-lift samples. The results are displayed in Table 17.

During inspection and sampling activities, visible mold was observed on a hallway wall, behind vinyl wallpaper, and under several window air conditioners. Sampling results showed that eight of the twelve areas tested had some level of mold growth. Three areas were heavily infested with toxic species at levels heavy enough to be of concern. *Stachybotrys chartarum*, reported to be a cause or contributing factor in memory loss, was identified in the Alzheimer's ward of this facility.

> **The dangers of mold exposure to our senior citizens and to our youngest citizens, our children, are far greater than they are to normally healthy adults. And like our children, our seniors are often dependent on others to be their voice and to assure their safety.**

We recommended diffusing with EOB2, cleaning visible mold with the cleaner containing EOB2, removal of all water-damaged and mold-infested materials, and other measures to eliminate excess moisture. Facility administrators decided to send in-house staff for training in mold remediation.

Observations and Conclusions

Although the administrators of this facility are to be applauded for sending their maintenance people for education in mold-remediation procedures, this case illustrates the reluctance of administrators of health-care facilities, and of many individuals, to go beyond established policies and procedures.

Misinformation regarding the use of bleach and disinfectants to treat mold has persisted for years, even though we know they do not work. Having ample evidence that conventional mold-remediation methods do not work and that the EOB2 protocol does work, we believe the administrators' decision is most unfortunate.

It is not unusual to find conditions conducive to mold growth in senior citizens' apartments, assisted-care facilities, and 24-hour care facilities. Occupants may be physically or mentally disabled, and spillage of food and drink, as well as unreported leaks and seepage from equipment, often provide sufficient moisture for mold growth.

The dangers of mold exposure to our senior citizens and to our youngest citizens, our children, are far greater than they are to normally healthy adults. And like our children, our seniors are often dependent on others to be their voice and to assure their safety.

TABLE 17 - CASE STUDY NO. 17

Mold Species	Office	Area A	Unit 3	Unit 6	Area B	Kitchen	Unit 4	Unit 5	O/S	Res. 3	Din Rm	Hall	Res. 4
Alternaria	7	0	0	0	0	0	0	0	47				
Ascospores	227	1040	747	140	500	160	87	20	8000				
Basidiospores	933	4213	1733	707	980	587	287	107	7200				
Bipolaris/Drechslera	7	7	0	0	7	0	7	0	7				
Botrytis	0	0	**20**	0	**520**	0	0	0	0				
Chaetomium	0	0	0	0	0	0	0	0	0	Heavy	Light	Light	V Hvy
Cladosporium	160	440	453	40	0	47	73	100	5333	Light	Light		
Curvularia	7	0	0	0	0	0	0	0	7				
Epicoccum	0	0	7	0	0	0	0	0	27				
Other Colorless	20	13	7	7	20	20	13	7	60				
Other Brown	0	0	0	0	0	0	7	0	0				
Penicillium/Aspergillus	93	120	333	40	93	40	20	13	107	Mod	Light	Light	Heavy
Rusts	0	0	0	0	7	0	0	0	0				
Smuts, etc.	13	33	0	0	73	0	27	7	113				
Stachybotrys	**7**	0	**13**	0	0	0	0	0	0	V Hvy		Light	V Hvy
Stemphylium	0	0	0	0	0	0	0	0	0			Light	Light
Torula	0	0	0	0	0	0	0	0	7				
Cercospora	0	0	0	0	0	0	0	0	33				
Oidium	0	0	0	0	0	0	0	0	13				
Exophiala	0	0	0	0	0	0	0	0	0			Heavy	
Ulocladium	0	0	0	0	0	0	0	0	0			Light	
Total	1474	5866	3313	934	2200	854	521	254	20954	G	G	G	G

Notes: The last four columns are tape-lift results **Bold** = Mold Growth in the area tested

O/S = Outside; **Mod** = Moderate; **V Hvy** = Very Heavy; **G** = Growth in area tested

Chapter 4

CASE STUDY NO. 18 – Private Residence with Small Children

The owner of an older rental house asked us to investigate the property when the occupants, a young mother with two small children, moved out because they were having constant headaches, colds, and flu-like symptoms. The young family also reportedly smelled musty odors in the house.

Inspecting to determine the source of the odors, evidence of rodent activity in the basement, crawl spaces, and HVAC ducts was found. We also found black mold growing on the walls and floor of the basement. Air samples and tape-lift samples were collected. The results of the sampling efforts are displayed in Table 18.

The levels of mold growth and mold-spore concentrations found inside the house posed the potential for health threats, especially to small children. The owner had a pest-control company take care of the rodent problem. Following our recommendations, the basement was cleaned, and then EOB2 was diffused continuously for 24 hours in the basement and the living areas.

Observations and Conclusions

Twelve species of mold were found inside the house, including five of them growing inside. Of these species, *Stachybotrys, Penicillium/Aspergillus, Cladosporium,* and *Chaetomium* are reported to cause and/or aggravate a large variety of health problems.

The lab analysis report shows thirty-three *Stachybotrys* spores per cubic meter in the outdoor air sample. Because *Stachybotrys* is a heavy, sticky spore, finding it in an outdoor air sample is unusual. As a result of this unusual report, sample-collection procedures

were revisited for this site. It was noted that the outdoor air sample had been collected with equipment placed atop a wooden deck that extended twenty-five feet from the back of the house. Returning to the site, it was also discovered that a crawlspace vent was located under the wooden deck where the sample had been taken, and the crawl space was open to the basement where a heavy growth of *Stachybotrys chartarum* had been found.

This case study underlines the importance of collecting outside air samples at least fifty feet away from any building being inspected.

Photo 9: Various Species of Mold on Concrete Basement Wall.

The young family moved back into the house and has continued diffusing regularly as a preventative measure. The owner has reported that they have not experienced any further problems with headaches or unusually severe or chronic illnesses. The owner and the family are both happy with the results.

No post-treatment samples were collected at this site.

Chapter 4

TABLE 18 - CASE STUDY NO. 18

Mold Species	Inside	Outside	Ratio	Tape 1	Tape 2
Acremonium	0	0	NA		Moderate
Alternaria	33	393	0.08		
Ascospores	720	4890	0.15		
Basidiospores	787	6140	0.13		
Bipolaris/Drechslera	7	67	0.1		
Chaetomuim	**13**	0	**GIO**		Light
Cladosporium	2370	22600	0.11	Heavy	
Curvularia	33	40	0.83		
Epicoccum	0	67	0.00		
Fusarium	20	27	0,74		
Nigrospora	0	107	0.00		
Penicillium/Aspergillus	**23400**	960	**24.4**	Moderate	Light
Pithomyces	0	47	0.00		
Smuts, etc.	13	160	0.08		
Stachybotrys	7	33	**0.21**		Heavy
Torula	0	27	0.00		
Cercospora	7	1120	0,01		
Total	27410	36678	0.75	Growth	Growth

Notes: Wall & floor samples are tape-lift samples from the basement

BOLD = Species growing inside

GIO indicates Growing Inside Only

NA = Not Applicable

CASE STUDY NO. 19 – Civil War Era Farm Home

The owner of a large, historic, two-story, brick farm home, built just after the Civil War, decided to turn his property into a bed-and-breakfast. Because of the age of the house, its setting in a grove of very old trees, a musty odor, and some gray spots on window frames, he decided to have it checked for mold. Mold growth on the window frames was visually identified as *Cladosporium*. One spore trap air sample was collected on the ground floor in the central hallway and one outside for comparison. The results are displayed in Table 19.

> This is an example of an older house with nearly 7,000 mold spores per cubic meter inside, yet the occupant indicated he had suffered no known adverse health effects.

Surprisingly, the samples did not find *Stachybotrys* spores present at this site. This building did, however, have the highest number of mold species growing inside the house, relative to those species found outside, of all the Case Studies presented in this book. Twenty-two species were identified in the air samples. Eight species were only found inside. Sixteen species, or 73% of the species identified, had substantially higher concentration levels inside than outside,

indicating these species were growing inside the house. Two other species had inside concentrations that were at levels equal to outside concentrations. It is possible that some growth of these species was occurring inside the house as well.

The owner purchased diffusers and the essential oil blend, diffused for 24 hours continuously, and then put the diffusers on timers so they would automatically run every 4 hours for 15 minutes. There has been no further sampling or other follow up at this time.

Observations and Conclusions

This is an example of an older house with nearly 7,000 mold spores per cubic meter inside, yet the occupant indicated he had suffered no known adverse health effects. This may be because the owner does not live in the main part of the house. He lives in the summer kitchen off the back porch, isolated from the house.

The owner continues to diffuse EOB2 in the house on a regular basis and reports that he enjoys the smell, as do his visitors and guests.

116

TABLE 19 - CASE STUDY NO. 19

Mold Species	Inside	Outside	Ci/Co
Alternaria	107	120	0.89
Ascospores	**80**	13	**6.15**
Basidiospores	**160**	27	**5.93**
Bipolaris/Drechslera	33	33	1.00
Chaetomuim	7	7	1.00
Cladosporium	**3467**	2667	**1.3**
Curvularia	**80**	7	**11.43**
Epicoccum	160	240	0.66
Fusarium	**7**	0	**GOI**
Nigrospora	**20**	0	**GOI**
Other Colorless	0	53	0.00
Other Brown	**87**	0	**GOI**
Penicillium/Aspergillus	**2267**	320	**7.08**
Rusts	**20**	7	**2.86**
Smuts, etc.	267	40	6.68
Stemphylium	0	7	0.00
Torula	**13**	0	**GOI**
Ulocladium	**7**	0	**GOI**
Scopulariopsis	**7**	0	**GOI**
Pithomyces	**40**	7	**5.71**
Spegazzinia	**7**	0	**GOI**
Pyricularia	**13**	0	**GOI**
Total	6849	3548	1.93

Ci = Concentration (Mold Spores per cubic meter) Inside

Co = Concentration Outside

Bold indicates growth inside

GOI = Growth Only Inside

Chapter 4

CASE STUDY NO. 20 – Private Residence in Bollinger County, Missouri

A woman in her sixties and her ninety-year-old mother were living in the family home, a house built in 1926. As the mother's health got progressively worse, conditions in the home also deteriorated. The situation became more and more stressful, and the daughter's health began to suffer. Increasingly, she had memory lapses and found it hard to express herself. She was often confused. The family doctor told her it was just ageing and stress, and prescribed vitamins and antidepressant drugs, which did not help.

When her mother passed away, relatives came for the funeral and stayed to visit for a few days. They were appalled at the conditions in the house and at the daughter's state of mind. A younger sister decided to stay and help clean up the house. After spending one night in the house, the sister began to experience severe headaches. After the second night, she had a red and swollen eye. She noticed dark splotches of mold growing in the basement and on the windows of the bedrooms.
.

The sister began researching the effects of mold and contacted a local friend who called us to do a mold inspection. Because of concern for their health and familiarity with essential oils, they began diffusing EOB2 before the inspection and sampling could be undertaken. Upon arrival at the site, the older sister could not form complete sentences during discussions with her.

Mold was found growing in multiple locations, and excess moisture was found in the basement from a plumbing leak in the sewer line and a crack in the basement wall. Air samples were collected outside and inside from the main floor and the basement, and three tape-lift samples were taken from surfaces in the bedrooms and basement. The results are displayed in Table 20.

Observations and Conclusions

There were at least six species, *Chaetomium, Cladosporium, Nigrospora, Penicillium/Aspergillus, Stachybotrys*, and *Ulocladium*, were growing inside the home. From experience at other sites where EOB2 was diffused, we know that the spore counts would have been higher before diffusing began. We recommended continuation of diffusing, that they move out of the house until repairs and remediation could be completed, and that they have professionals repair leaks and remove mold- and water-damaged materials.

The two sisters moved out of the house and returned to the younger sister's home in another state. The status of cleanup, remediation, and repair is unknown at this time; however, we did hear through the local friend that within a month of moving out of the mold-infested home, while continuing a program of diffusing and taking supplements, the older sister's health had improved dramatically.

TABLE 20- CASE STUDY NO. 20

Mold Species	Outside	Bedrm.	Ratio	Basemnt	Ratio	Tape 1	Tape 2	Tape 3	Light
Alternaria	73	0	0	27	0.37				
Ascospores	3247	87	0.03	100	0.03				
Basidiospores	11560	140	0.01	307	0.03				
Chaetomuim	0	7	**GIO**	7	**GIO**				
Cladosporium	5193	333	0.06	727	0.14	Growth	Growth	Growth	
Curvularia	20	20	1.00	7	0.35				
Epicoccum	0	0	NA	7	**GIO**				
Nigrospora	0	7	**GIO**	7	**GIO**				
Other Colorless	213	60	0.28	0	0.00				
Other Brown	7	7	1.00	7	1.00				
Penicillium/Aspergillus	560	387	0.69	**2700**	**4.82**				
Pithomyces	7	7	1.00	0	0.00				
Smuts, etc.	567	40	0.07	120	0.21				
Stachybotrys	0	13	**GIO**	7	**GIO**				
Ulocladium	0	0	NA	20	**GIO**				
Stemphylium	7	0	0.00	0	0.00				
Polythrincium	7	0	0.00	0	0.00				
Total	21461	1108	0.05	4043	0.19	**Mod.**	**Mod.**	**Mod.**	**Light**

Notes: Tape 1 and Tape 2 are tape-lift samples from bwdroom windows, Tape 3 is from the basement.

BOLD = Species growing inside   NA = Not Applicable

GIO indicates Growing Inside Only   **Mod.** = Moderate Growth

Testing the Hypothesis

We have looked at the background information, the sampling, the treatment, and the results for each of twenty individual case studies. These data, gathered and documented by a strict chain of custody protocol and analyzed by qualified microbiologists in a certified environmental laboratory, provide the scientific basis required to answer the question of whether the blend of essential oils (EOB2) used in these case studies was effective in the elimination of mold in general and toxic mold in particular.

The scientific method is defined as proposing a hypothesis, designing experiments to test the hypothesis, and analyzing the results of the experiments to determine whether the hypothesis is true or false. When data sets are collected from a number of different sites, in addition to validating or negating a hypothesis, it also becomes possible to conduct a secondary level of analysis. From the data, we are able to gain additional information about the subject under investigation and to determine the overall effectiveness of the methods used.

Studying individual cases provides insights, and when the results of all the studies are combined, more information can be gleaned from the data. In this instance, in addition to determining whether this particular blend of essential oils is effective in eliminating indoor mold, the case-study data also provide valuable information about the frequency of occurrence of individual mold species in the area where the case studies were conducted, the relative occurrence of various mold species in buildings, overall effectiveness of the protocol utilized, and effectiveness against individual species of mold.

In each case study, we documented the protocol used and determined the spore-removal efficiency (SRE) for individual mold species. We gradually improved the protocol as we learned more from each case study. It evolved from simply diffusing the blend for 24 hours in conjunction with conventional cleaning and removal of damaged building materials to a more sophisticated approach involving diffusing, using the EOB2 based household cleaner, and tailoring diffusing time, cleanup and removal to site-specific conditions and circumstances. The latest version of the protocol is presented in Chapter 7.

By combining the information from all the case studies, we can answer the following questions:

1. Which mold species were most commonly found in all samples collected, including both indoor and outside samples?

2. How much variation in mold species present did we find from site to site?

3. Which mold species were most commonly found in indoor samples collected, and what was the average number of species found in indoor samples?

4. What was the average number of mold species found in outdoor samples?

5. Which species of mold were most commonly found in the outside air?

6. What was the average number of species found growing inside the case-study buildings?

7. Which mold species were most commonly found growing inside the case-study buildings?

8. How effective were the EOB2 protocols on the various mold species and overall?

9. Was the treatment equally effective in all cases?

10. Was the protocol effective against all mold species encountered?

11. Which mold species is the protocol most effective against?

12. Which mold species is the protocol least effective against?

13. How does EOB2 compare with other mold-treatment products?

In order to answer these questions, we have ranked the information from all the case studies according to frequency of occurrence,

frequency of indoor growth, and average SRE values. These rankings are presented in the Tables 21, 22, and 23. Referring to these tables, as well as individual case-study tables, and field notes, we can answer the questions posed above.

Question 1: Which mold species were most commonly found in all samples collected, including both indoor and outside samples?

Answer: Referring to Table 21, we see that the mold-spore classification *Penicillium/Aspergillus* was most common. *Basidiospores and Cladosporium*, while not quite as common as *Penicillium/Aspergillus* spores, were also found at all twenty sites.

Note: When viable samples are collected, laboratory analyses distinguish between *Aspergillus* and *Penicillium*. In cases where only spore-trap air samples are collected, laboratory analyses do not distinguish between these two species.

Question 2: How much variation in mold species present did we find from site to site?

Answer: Considerable variation in the number and type of mold species was found from site to site and even from room to room at the same site. The number of individual species at case-study sites ranged from as few as twelve to as many as twenty-eight, and the mix of species varied from site to site, with forty-two different species and types of molds represented in the twenty case studies.

Question 3: Which mold species were most commonly found in indoor samples collected, and what was the average number of species found in indoor samples?

Answer: The average number of species found in indoor samples was twelve species, with *Cladosporium* and *Penicillium/Aspergillus* found inside at all twenty sites.

Question 4: What was the average number of mold species found in outdoor samples?

Answer: The average number of species in outdoor samples was eleven species.

Question 5: Which species of mold were most commonly found in the outside air?

Answer: *Basidiospores*, *Cladosporium*, and *Penicillium/Aspergillus* were found in outdoor samples at all sites.

Question 6: What was the average number of species found growing inside the Case Study buildings?

Answer: Eight species.

Note: Finding spores in indoor air samples does not necessarily mean that those species are growing inside, because there is always an exchange of air between inside and outside. Identification of species growing inside case-study buildings was done by one or more of the following means: laboratory analyses of viable samples, comparison of indoor and outdoor concentrations, and visual identification.

Question 7: Which mold species were most commonly found growing inside the case-study buildings?

Answer: Referring to Table 22, we see that *Stachybotrys* was found growing inside seventeen of the twenty sites.

Chaetomium was found growing in sixteen, and *Penicillium* and/or *Aspergillus* was found growing in fifteen of the case-study buildings. All the rest were found growing in fewer than ten sites, and six of the species found in air samples were not found growing in any of the case study buildings.

Question 8: How effective were the EOB2 protocols on the various mold species and overall?

Answer: Looking at Table 23, we see that the protocol was 100% effective against seventeen of the mold species found in the twenty case studies, and the effectiveness of the protocol overall (i.e., the average of the averages for all species) was 96.65%.

Question 9: Was the treatment equally effective in all cases?

Answer: No, but the range of effectiveness is narrow. Overall spore removal efficiency ranged from 93% for Case Study No. 13, to 99.8% for Case Study No. 11, with an average of 96.7%, and a standard deviation of 2.48.

Question 10: Was the protocol effective against all mold species encountered?

Answer: Yes. The protocol was 90% to 100% effective against all of the mold species of concern, and 100% effective against toxin-producing and infection-causing species, including *Stachybotrys*, *Bipolaris/Drechslera*, and *Aspergillus*, in most of the case studies.

Question 11: Which mold species is it most effective against?

Answer: The protocol was 100% effective against seventeen of the mold species found in the case studies, including *Nigrospora*, *Rusts*, *Spegazzinia*, *Cercospora*, *Fusicladium*, *Tetraploa*, and others, many of which produce mycotoxins harmful to humans. The average efficiency against *Stachybotrys* was 99.78%.

Question 12: Which mold species is the protocol least effective against?

Answer: The lowest average spore-removal efficiency for any species found in these Case Studies was for *Basidiospores*, at 85.5%. However, the SRE for *Basidiospores* was 100% in two of the case studies.

Question 13: How does EOB2 compare with other mold-treatment products?

Answer: Dr. Richard J. Shaughnessy, PhD, Program Director for the University of Tulsa Indoor Air Program in Tulsa, OK (in an article that appeared in the June, 2007, issue of "Indoor Environment Connections, The Newspaper for the IAQ Industry,") said:

"There are many... examples of IAQ products that 1) have little or no significant benefit to the control of indoor contaminants, and, most egregiously, 2) result in an actual detriment to the health of the poorly advised consumer."

Based on the case-study data presented, when EOB2 was used in accordance with the Close Essential Oil Protocol, it demonstrated an ability to provide benefit to the control of indoor contaminants, benefit to the health of the consumer, and was found to be superior to other mold treatment products currently available. Laboratory test data for essential oils applied to specific fungi growth in petri dishes are referenced in Chapter 3, "Show Me the Science," and direct comparison of the results of using bleach and chemical fungicides with the results of diffusing essential oils is discussed in Chapter 4, "Case Studies."

A one-to-one comparison of the efficacy of different mold-eradication methods (including the 10-Step Protocol using the Thieves® essential oil blend) would require applying each method to identical mold infestations and determining the effectiveness of each by collecting before and after data. Direct, one-to-one comparison with other mold-eradication methodologies was not undertaken as a part of this study but is suggested as additional research. What follows is an objective comparison of the Close Essential Oil Protocol with other mold-remediation methodologies, based on available information.

The criteria used are:

Mold eradication
Mold prevention
Health effects
Residual effects

Numerical data for these criteria are not available for all methods. Therefore, scoring is based on ranking each method as follows:

3 = objective evidence of excellent results
2 = objective evidence of good results
1 = objective evidence of minor positive results

0 = no objective evidence of positive or negative results

-1 = objective evidence of minor negative results

-2 = objective evidence of significant negative results

-3 = objective evidence of serious negative results

Notes on Criteria:

A. Mold Eradication

Treatments that are known to remove mold spores from the air and kill spores and source colonies were awarded a 3. Treatments that kill mold by limited-area application (spot treatment) or air-stream treatment were given a 2. If a treatment kills only some mold species and/or retards their growth, it was given a 1.

B. Mold Prevention

If a treatment prevented mold growth for at least one week or more, when applied in an environment conducive to mold growth, it was awarded a 3. If prevention was localized, as with ultraviolet (UV) treatment, a 2 was assigned. If growth was retarded, a 1 was given.

C. Health Effects

If positive health effects were reported, as in several of the case studies in this book — especially if corroborated by medical exams — a 3 was awarded. If positive health effects were noted in most cases, a 2 was awarded. If positive health effects were rare but reported, a 1 was awarded. If no health effects, positive or negative, were reported, a 0 was assigned. If minor adverse health effects were noted but could be avoided by careful application of the treatment or by taking appropriate protective measures, then a -1 was given. If major adverse health effects were possible, even if they could be avoided by taking appropriate

protective measures, a -2 was assigned. And if major adverse health effects were possible in spite of taking protective measures and using appropriate controls, a -3 was assigned.

D. Residual Effects

If a treatment had documented residual effects, high mold prevention, and high positive health effects, a 3 was awarded. If residual effects were noted for a treatment with positive health effects, and preventative power, a 2 or 1 was awarded, depending on the combined health and prevention score. If there was no residual effect, a 0 was assigned. If there were residual effects with negative health effects or damage to materials, a -1, -2, or -3 was assigned, depending on the level of the negative health score.

Individual scoring and total score comparisons are presented in Table 24.

It is recognized that there are variations in treatment protocols and product contents within treatment categories, especially chemical fungicides, oxidizers, and disinfectants, so that some specific products and treatment will perform better than others. The scores represent each category as a whole.

HYPOTHESIS:

The use of the essential oil blend EOB2 is effective in elimination and control of mold inside buildings.

CONCLUSION:

The hypothesis is confirmed. The essential oil blend used in these studies was very effective on all species encountered, including toxic mold species.

Suggestions for Additional Research

Analysis of the data collected from the case studies reported in this book has answered many questions. However, as might be expected by anyone familiar with or involved with research, it has also raised additional questions that suggest ideas and directions for additional research. Following are some suggestions that we believe would further the knowledge and understanding of the potential benefits essential oils offer for the remediation of mold.

1. Clients have asked, "If diffusing EOB2 kills mold spores and removes them from the air, how does it do that, and where do the spores go?"

These are two important questions, especially since we know that dead mold spores can cause health problems. The data suggest that the spores are broken down or digested by the essential oils, with the possible exception of *Penicillium*. This hypothesis can be tested by applying EOB2 to cultures of the mold species of concern and observing them under a microscope.

2. EOB2 is the only blend of essential oils in the Case Studies for which data was collected. Could there be other blends that would be just as effective or perhaps more effective under certain conditions?

> **To our knowledge, EOB2 (Thieves® essential oil blend) is the only mold-remediation or mold-treatment option available that can be diffused as a preventative in the home or any building for extended periods of time and while occupants are present.**

Laboratory studies demonstrate the antifungal properties of numerous species of Thyme, Oregano, Cinnamon, Citronella, Mint, Sage, Eucalyptus, Lemongrass, Tea Tree, and Clove essential oils (for a limited list, see Chapter 3, "Show Me the Science"). The common names for these oils are given above, however, the scientific name is preferred for studies because each species has different chemical components.[1-9]

A single plant species typically contains 100 to 400 different chemical compounds. Plants are classified into broad categories called families, such as *Lamiaceae* or *Labiatae* for the Mint family, and *Pinaceae* for the Pine family. Families are then differentiated by genus (there are numerous genuses within a family), and by species. There may be numerous species within a genus.

For example, Tea Tree oil is *Melaleuca alternifolia*. Cajuput oil is *Melaleuca cajuputi*. Niaouli, Nerolina, and MQV are all common names for *Melaleuca quinquenervia*.

Thyme oil is usually *Thymus vulgaris* CT *linalol* (also known as Garden Thyme), or *Thymus vulgaris* CT *thymol* (also known as Red Thyme).

The CT in these Latin identifiers stands for *chemotype* which indicates that the specific plant came from the same genus, the same species, and the same seed; however, the essential oil has a specific, dominant chemical constituent, and may come from a specific region of the world. Other species of Thyme essential oil include:

> Moroccan Thyme (*Thymus satureioides*)
> Spanish Marjoram (*Thymus mastichina*)
> Sweet Thyme (*Thymus vulgaris* CT *thujanol*)

Understanding this, the importance of identifying the specific plant species and chemotype (when appropriate) becomes clear.

Generally, the best blends, or balanced synergies, of essential oils contain:

50 to 60% phenols or ketones (Top
Notes)
20 to 40% monoterpenes (Middle
Notes)
10 to 20% sesquiterpenes (Base
Notes)

If you selected only eight specific
species/chemotypes of essential oils known to
have antifungal properties and combined only
three at a time in a well-balanced blend, you
would have fifty-six different combinations to
be tested.

Mathematically, there are 247 possible
combinations of eight essential oils. It took us
about twenty months to produce the data
from twenty case studies with EOB2 (EOB2
is a combination of five specific species of
essential oils formulated by the manufacturer,
Young Living Essential Oils.). At that rate, it
would take over 400 years to field test the
possible combinations available with only
eight specific species/chemotypes of essential
oils.

Comparing different blends, or synergies, of
essential oils that are known to have
antifungal properties presents almost endless
opportunities for research. Fortunately, the
field tests and case studies presented in this
book provide ample evidence that EOB2 (the
Thieves® essential oil blend) is a good one to
choose first.

3. We have found in the field tests
conducted that using the Thieves®
Household Cleaner, undiluted, offers the
least possibility of providing additional
water that could lead to a rebound of
mold growth. We also found it beneficial
to add an extra 15 ml bottle of EOB2
(Thieves® essential oil blend) when using
the cleaner to spray crawlspaces or large
areas infested with mold. Additional
research and comparison tests are
recommended to determine whether
strengthening the cleaner with additional

EOB2, Oregano, or other essential oils
would provide better results.

4. The diffusers used in the case studies were
ones that were readily available to us.
Research is needed to determine the
optimal design for diffusers, based on the
building in which it is being used.

The diffusers used in the case studies were
designed for dispersion of essential oils in a
space of about 1,000 sq. ft., with an average
floor to ceiling height of 8 to 10 feet. They
were not designed specifically for mold
eradication. The average oil output with the
available diffusers was found to be about 10
ml per day, or a little more than 0.4 ml per
hour. For large buildings, such as
gymnasiums, libraries, warehouses, schools,
hospitals, and large office buildings, we placed
one or more diffusers in air-handling units,
and in individual rooms and offices, and at
points within the large spaces that would
maximize the amount of oil dispersed in a
limited time frame. Experience suggests that
the diffusers used worked quite well.
However, diffusers with more power, capable
of diffusing at least 1 ml per hour or more
might be more effective, especially in large
spaces.

5. Additional research is needed to find the
most effective ways to apply EOB2
and/or other essential oil blends to mold
infested spaces and areas.

The different methods of applying EOB2
used during the case studies included:
diffusing, direct application of EOB2, and/or
the Thieves® Household Cleaner, using
sponges or paper towels, and also spraying the
Thieves® Household Cleaner into difficult-
to-reach spaces (such as crawl spaces) with
modified paint sprayers and agricultural
pesticide sprayers. Again, the methods proved
effective. However, cost-benefit analyses for
the various methods could be additional areas
for research.

6. One of the most challenging problems in evaluating the effectiveness of treatment of indoor areas and spaces is determining the rate of indoor-outdoor air exchange and the resulting effects on the concentrations of various species of mold spores. Research and development of methods and equipment to measure the indoor-outdoor exchange of mold spores is much needed.

7. Standardization and documentation of health problems occupants experienced in buildings both before and after treatment with the 10-Step EOB2 Protocol would aid in identifying links between health and mold exposure.

8. Field tests using food-grade, perfume-grade, and synthetic essential oils in the 10-Step Protocol, instead of EOB2, could be conducted to determine their effectiveness when compared with EOB2. Standardized documentation of health problems experienced both before and after treatment with these lower-grade essential oils would provide helpful information for consumers.

9. We know that molds are adaptable. Long-term tests using a variety of essential oils are needed to determine whether mold species can adapt to essential oils and which ones they can adapt to, if any. A related question is: Do plants adapt also by varying the makeup of their essential oils?

An essential oil's chemical makeup varies slightly from year to year, even when the plant is grown from the same seed and in the same field. Essential oils are often compared to wines because this variance may enhance or diminish the rich bouquet of chemical compounds derived from a particular oil distilled in a particular year. The changes in chemical makeup result from varying amounts of rainfall, sunlight, temperature, and

variations in the chemistry of the soil, the time of harvest, and the distillation process. So, it is quite possible that mold species would not be able to adapt to organically grown and bottled, therapeutic grade essential oils. However, this is not known with certainty and, therefore, offers opportunity for research.

In summary, our recommendations for additional research include:

1. Tests to determine what happens to mold spores treated with EOB2.

2. Laboratory and field tests of other promising blends of essential oils.

3. Research and testing to improve the effectiveness of the Thieves® Household Cleaner on mold.

4. Research and development to improve the efficiency and effectiveness of diffusers.

5. Research and development to design effective equipment for applying the Thieves® Household Cleaner to crawlspaces, HVAC systems, and other hard-to-reach spaces.

6. Research and development to design methods and equipment to measure indoor/outdoor air/mold spore exchange.

7. Standardized determinations for assessing and documenting an occupant's health before and after treatment.

8. Comparison of lower grade essential oils with EOB2.

9. Tests to determine whether mold species adapt to essential oils.

Things To Remember

In addition to confirming the hypothesis, another unexpected positive result was discovered. The data show that diffusing EOB2 continued to be effective in keeping spores absent from air samples up to several months, even when diffusing was discontinued.

EOB2 is superior to conventional mold-remediation treatments in several ways:

1. In contrast with toxic fungicides, including bleach, EOB2 is non-toxic and is approved for human consumption by the FDA.

2. Studies indicate that EOB2 (the Thieves® essential oil blend) and some of the individual or single essential oils included in this blend are antibacterial,[1-9] antiseptic,[1-9] and antiviral.[1-9] The manufacturer of this blend states that it may be beneficial to human health. Our personal experiences with this blend have proven to us that it is very beneficial to human health.

3. Three of the essential oils in this blend (Cinnamon, Clove, and Rosemary) are specifically exempted from the Federal Insecticide, Fungicide and Rodenticide Act (FIFRA).[10] Lemon oil is no longer registered under FIFRA.[11] Eucalyptus oil received an emergency exemption from FIFRA in 2003 for use in or on honey and honeycomb.[12]

4. EOB2 removes mold spores from the air. This blend of essential oils does not just kill mold spores, it actually removes mold spores, both living and dead, from the air.

5. Although additional research is required, the bulk samples collected provide strong evidence that the Close Protocol is effective for both porous and non-porous materials.

6. Case studies indicate this treatment provides a residual effect that inhibits future growth of mold for up to five months.

7. Case-study participants reported improvements in their health while following the Close Essential Oil Protocol.

8. Case-study participants reported enjoying the cinnamon and cloves smell of the essential oil blend.

9. EOB2 is much more user-friendly, eco-friendly, and pet-friendly than all other currently available treatments, and may be used on a daily basis as a preventative.

This blend of essential oils is non-toxic, FIFRA exempt,[10-12] and FDA approved for human consumption. It is antibacterial,[1-9] and antiviral,[1-9] and supports human health.[1-9] No other mold-treatment option available can say this. No other fungicide, bleach, or mold-remediation technology offers such a cost-effective means to eliminate and prevent mold.

No other treatment currently available provides this level of protection without costly or poisonous drawbacks.

To our knowledge, this is the only mold-remediation or mold-treatment option available that can be used as a preventative in the home or office building for long periods of time. The details of how to use this treatment regularly, as a preventative measure, are described in Chapter 7.

The original questions posed in the very beginning of this book were:

1. Is there a non-toxic solution to toxic mold?

2. Can essential oils be used to effectively remediate mold in buildings?

These questions can now both be answered with an unequivocal YES!

References

[1] Advanced Aromatherapy, The Science of Essential Oil Therapy, Kurt Schnaubelt, PhD, Healing Arts Press, 1998.

[2] Aromatherapy For Health Professionals, by Shirley Price and Len Price, Churchill Livingstone, 1995-2001.

[3] Clinical Aromatherapy, Essential Oils in Practice, Second Edition, Jane Buckle, RN, PhD, Churchill Livingstone, 2003

[4] Healing Oils Healing Hands, Linda L. Smith, RN, MS, CCA, HSTM Press, 2003.

[5] Integrated Aromatic Medicine, Proceedings from the First International Symposium, Grasse, France, March, 1998. Translated form the French and published by Essential Science Publishing, 2000.

[6] Medical Aromatherapy, Healing with Essential Oils, Kurt Schnaubelt, PhD, Frog, Ltd., North Atlantic Books, 1998.

[7] Screening for Inhibitory Activity of Essential Oils on Selected Bacteria, Fungi, and Viruses. Chao, SC; Young, DG, Oberg, CJ. Journal of Essential Oil Research, 1998.

[8] The Chemistry of Essential Oils Made Simple, God Manifest in Molecules, by David Stewart, PhD, DNM, CARE Publications, 2005.

[9] The Practice of Aromatherapy, Jean Valnet, MD, edited by Robert Tisserand, Healing Arts Press, 1990.

[10] Code of Federal Regulations, Title 40, Volume 22, Revised as of July 1, 2004, from the U.S. Government Printing Office via GPO Access, Cite: 40CFR152.25, Page 11-13.

[11] Code of Federal Regulations, Title 40, Part 180, Volume 71, Number 187, Revised September 27, 2006, from the Federal Register Online via GPO Access [wais.access.gpo.gov], DOCID: fr27se06-14, Page 56378-56383, "Bentazon, Carboxin, Dipropyl Isocinchomeronate, oil of Lemongrass (Oil of Lemon) and Oil of Orange; Tolerance Actions."

[12] Code of Federal Regulations, Title 40, Part 180, Volume 68, Number 109, June 6, 2003, from the Federal Register Online via GPO Access [wais.access.gpo.gov] DOCID: fr06jn-15, Page 33876-33882, "Thymol and Eucalyptus Oil; Exemptions from the Requirement of a Tolerance.".

TABLE 21 - COMPARISON OF EOB2
WITH OTHER MOLD TREATMENTS

Mold Treatment	Criteria				TOTAL SCORE
	A Eradication	B Prevention	C Health Effects	D Residual Effects	
EOB2	3	3	2	2	10
UV Light	2	2	1	0	5
Disinfectant	0	1	2	1	4
Oxidizers	3	2	-2	-2	1
Chemical Fungicides	3	2	-3	-1	1
Bleach	2	1	-2	0	1

TABLE 22 - SPECIES RANKED BY FREQUENCY OF OCCURRENCE

Mold Species	Cib	Cob	Cia	Coa	Total
Penicillium/Aspergillus	34	20	13	13	80
Cladosporium	32	20	13	13	78
Basidiospores	32	20	12	13	77
Smuts, etc.	22	17	11	12	62
Ascospores	25	17	9	11	62
Other Brown	19	12	12	8	51
Alternaria	17	18	6	9	50
Other Colorless	17	15	6	9	47
Epicoccum	17	14	4	8	43
Chaetomium	17	5	6	4	32
Stachybotrys	24	3	2	0	29
Curvularia	14	9	2	2	27
Pithomyces	13	7	1	3	24
Bipolaris/Drechslera	12	8	0	3	23
Torula	7	8	1	4	20
Nigrospora	10	5	0	4	19
Fusarium	6	4	1	3	14
Rusts	7	5	0	2	14
Cercospora	3	4	0	3	10
Scopulariopsis	4	2	0	3	9
Spegazzinia	2	4	0	0	6
Stemphylium	3	3	0	0	6
Ulocladium	4	1	0	1	6
Oidium	2	2	0	1	5
Pestalotiopsis	3	1	0	1	5
Botrytis	3	1	0	0	4
Gliomastix	0	2	0	1	3
Peronospora	0	2	0	1	3
Tetraploa	1	2	0	0	3
Fusicladium	1	1	0	0	2
Polythrincium	1	1	0	0	2
Pyricularia	1	1	0	0	2
Trichoderma	1	1	0	0	2
Trichothecium	1	1	0	0	2
Acremonium	1	0	0	0	1
Arthrinium	0	1	0	0	1
Exophiala	1	0	0	0	1
Fusirella	0	1	0	0	1
Zygomycetes	1	0	0	0	1
TOTALS	358	238	99	132	827

TABLE 23 - SPECIES RANKED BY FREQUENCY
OF GROWTH INSIDE BUILDINGS

Mold Species	No. of Samples
Stachybotrys	17
Chaetomium	16
Penicillium/Aspergillus	15
Cladosporium	9
Smuts, etc.	9
Ascospores	8
Epicoccum	7
Other Brown	7
Curvularia	7
Nigrospora	7
Basidiospores	6
Pithomyces	5
Bipolaris/Drechslera	5
Torula	5
Alternaria	4
Other Colorless	4
Rusts	4
Scopulariopsis	3
Ulocladium	3
Spegazzinia	2
Stemphylium	2
Pyricularia	2
Fusarium	1
Cercospora	1
Pestalotiopsis	1
Botrytis	1
Tetraploa	1
Trichoderma	1
Acremonium	1
Exophiala	1
Zygomycetes	1
Fusicladium	0
Gliomastix	0
Oidium	0
Peronospora	0
Polythrincium	0
Trichothecium	0

TABLE 24
AVERAGE SRE & NUMBER OF CASES WITH 100% SRE BY SPECIES

Mold Species	Average SRE	Number of Cases with SRE = 100%	Number of Cases with Species Present
Bipolaris/Drechslera	100%	7	7
Nigrospora	100%	6	6
Rusts	100%	4	4
Spegazzinia	100%	4	4
Cercospora	100%	3	3
Fusicladium	100%	1	1
Gliomastix	100%	1	1
Odium	100%	3	3
Pestalotiopsis	100%	2	2
Peronospora	100%	1	1
Polythrincium	100%	1	1
Pyricularia	100%	1	1
Scopulariopsis	100%	5	5
Stemphylium	100%	3	3
Tetraploa	100%	2	2
Trichoderma	100%	2	2
Ulocladium	100%	2	2
Stachybotrys	99.78%	10	13
Torula	99.46%	6	8
Other Colorless	98.65%	10	13
Curvularia	98.64%	6	7
Epicoccum	98.00%	9	11
Penicillium/Aspergillus	93.42%	0	13
Cladosporium	92.67%	0	13
Pithomyces	90.91%	8	9
Smuts, etc.	90.86%	5	13
Ascospores	90.13%	6	13
Alternaria	90.04%	6	11
Chaetomium	89.09%	4	10
Fusarium	88.96%	4	5
Other Brown	86.80%	3	13
Basidiospores	85.50%	2	13
Average SRE =	96.65%		
	Totals:	127	213

CHAPTER 6

PRACTICAL SAFETY
WITH ESSENTIAL OILS

Ancient Wisdom – Modern Miracles

Rose, cinnamon, peppermint, eucalyptus – the alluring scents, sweet aromas, and healing properties of aromatic plants, spices and essential oils have been written about by poets and historians, and recorded in medical texts the world over for thousands of years. From India to Greece, China to Egypt, Europe to America, the healing properties of numerous aromatic plants and their oils have been recorded for posterity.

Anthropological evidence suggests that our Neolithic ancestors (7,000 BC to 4000 BC) combined fragrant plants with the fatty oils of olive and sesame to create the first known ointments and salves. Dr. Paolo Rovesti discovered a terra-cotta apparatus (dated to 3,000 BC) that is believed to be the first known primitive still that may have been used for distilling essential oils from plants.

The earliest known mention of an aromatic oil is found in the Sanskrit teachings that form the basis of India's Ayurvedic Medicine, written more than 5,000 years ago. In the *Iliad*, Homer describes how Aphrodite anoints the body of Hector with Rose oil. Shen Nung's Herbal Book, circa 2700 BC, the oldest known medical book in China, provided information on the medicinal uses of more than 300 plants. And perhaps more familiar to most of us, here in the Western world, is the Bible story which tells us that Frankincense and Myrrh were given to the Christ child. According to Dr. David Stewart (*Healing Oils of the Bible*, 2001), over two-thirds of the books in the Bible contain references to oils, aromatic plants, or anointing with oil, and anointing with oil for the purpose of healing the sick.

So why do we, here in the West, think aromatherapy and essential oils are something new?

While essential oils were used throughout history, in modern times there is one man who is often referred to the father of Aromatherapy: Rene-Maurice Gattefosse. Gattefosse, a Frenchman, re-discovered the healing properties of Lavender oil after severely burning his arm and hand in a laboratory explosion in July, 1910. According to his report of the incident, he stuck his hand into a vat of what he thought was water. To his amazement his pain ceased.

Gattefosse was a chemist, working in the laboratory at his family's perfumery business, and he had actually put his arm into a vat of pure Lavender essential oil. He reported that he applied the essential oil frequently to his "gangrenous burn" and it healed without any scar. This experience led him into a lifetime of research into the therapeutic benefits of essential oils. He is credited with coining the term "Aromatherapie." And his remarkable book, *Gattefosse's Aromatherapie*, first published in 1937, was translated into English in 1992 (The C. W. Daniel Company Ltd., Reprinted 1995.). In the Bibliography of his book, Gattefosse cites more than 200 relevant reference works completed between 1680 and 1933. Robert Tisserand, in the "Editor's Introduction" to the English translation, says Gattefosse speaks of his pioneering efforts as the "dogged work of a chemist and perfumer who patiently endeavored to prove the efficacy of fragrant substances." Tisserand goes on to say, "There can be no doubt that he [Gattefosse] did more than anyone to lay the foundations for aromatherapy."

Gattefosse shared his research with Jean Valnet, MD, a French surgeon, who used essential oils on patients suffering battlefield injuries, during World War II. When his supply of antibiotics was exhausted, Valnet turned to essential oils, and the essential oils exerted a powerful effect in combating and counteracting infection. He was able to save the lives of many soldiers who might otherwise have died.

Dr. Valnet published most of his research in medical journals, but he also published articles and books in the lay press. His best known book, *The Practice of Aromatherapy*, was originally published in French in 1948, and simply titled *Aromatherapie*. The English translation (Healing Arts Press, 1990) was edited by Robert Tisserand.

Note that these two seminal works were both published in English for the first time in the 1990's, and both were edited by Robert Tisserand. Robert Tisserand was an English massage therapist, who learned about essential oils from a woman named Marguerite Maury. Maury was an Austrian biochemist, who married a French physician, and brought the use of essential oils to the massage and spa industry. She created many spas and did lectures throughout Europe. Tisserand, Maury's student, wrote and published the very first book in English on aromatherapy in the year 1977.

That is why we think essential oils and aromatherapy are new, because these were not terms we had heard in the United States before the 1970s. Tisserand, and others, including Jeanne Rose (an American), are largely responsible for bringing the terms "aromatherapy" to the general public during the 1970s. Their focus and their training were related to the use of essential oils in the spa industry in Europe. This explains why we, here in the U.S., think aromatherapy is something used primarily in massage, facials, soaps, perfumes, and candles. Over time, the various forms and uses for essential oils have led to a multitude of schools of thought related to the use of essential oils.

Four Schools of Thought

As should be clear, essential oils have many benefits beyond mold remediation. A person's first experience of essential oils can be quite dramatic. Jacqui had experienced chronic, seasonal allergies for many years; yet, on her first exposure to essential oils all her allergy symptoms disappeared in under three minutes. This prompted an intense interest in knowing everything about essential oils, and Jacqui began her studies in 1995, reading everything she could find. What she learned perplexed and confused her. Contradictions abounded in the available literature. Some books said a particular essential oil should never be used, while other books extolled the healing benefits of the very same essential oil. It was not until she understood the various schools of thought and the many issues of toxicity and other problems that resulted from the use of inferior grade essential oils, and synthetic and adulterated essential oils that the confusion began to clear and the numerous contradictions began to be understandable.

There are four pathways through which essential oils enter the body:

1. Respiratory system (nose and lungs)

2. Skin

3. Digestive tract

4. Absorbent tissues of body orifices, such as the mouth.

In the world of essential oils and aromatherapy there are also four quite different schools of thought about how essential are to be used. This material has been covered in previous books on aromatherapy, but this brief discussion will help those who are unfamiliar with this information.

The proponents of each school of thought often think their approach is the only valid approach, and in some cases they aggressively oppose using essential oils in any other way. Briefly, these are the four schools of thought:

> French
> German
> English
> American

The French were responsible for developing what is called "Modern Aromatherapy" or "Medical Aromatherapy," because the proponents and researchers who founded this school were primarily chemists and medical doctors. They stress the aggressive use of essential oils taken into the body through all four pathways. Daniel Penoel, MD, (a student of Dr. Jean Valnet) wrote the books *L'Aromatherapie Exactement* which is the first definitive medical textbook on the chemistry and clinical application of essential oils

(written with Pierre Franchomme, PhD) and *Natural Home Health Care Using Essential Oils* (written with his wife, Rose-Marie Penoel). Dr. Penoel documents countless recoveries that occurred for his patients who used essential oils and were healed, when allopathic (traditional western) medicine had failed to provide them assistance or hope.

Dr. Penoel says everyone must be "aromatic warriors." This typifies what can be called the French school of aromatherapy, where the emphasis is on therapy. Most adherents to the French school believe that anyone can use essential oils with a modicum of knowledge, and a few good reference books. This school of thought is very similar to that of the Americans.

The German school of aromatherapy emphasizes the inhalation of essential oils. Inhalation allows the small molecules of essential oils to be absorbed by olfactory nerves, the brain, and the lungs where they enter the blood stream and spread throughout the body. This is a very effective means of enjoying the benefits of essential oils. The emphasis in the German school is on aroma and its therapeutic benefits. Kurt Schnaubelt, PhD, has written many books that are available in English (see the Bibliography) and has an excellent school, the Pacific Institute of Aromatherapy, available to those interested in learning more about essential oils.

This is a very safe way to enjoy essential oils, and adherents to the German school of thought believe anyone can and should enjoy the many benefits of aromatherapy.

The English, or British, school of aromatherapy is focused primarily on the use of essential oils with massage. Robert Tisserand, the British massage therapist who eventually opened his own company distributing essential oils in Europe and America, authored, co-authored, and/or edited many of the books that were originally available in England and America. His contribution to the field is substantial. However, his emphasis on the use of essential oils for massage and his concerns about the safety of essential oils have led to the popular notion among many trained in the British school that essential oils should never be used any other way.

The emphasis in the British system is on applying oils to the skin in a highly dilute form (2-5% essential oil diluted with 95-98% base of neutral carrier oil), in order to promote relaxation. This approach is definitely a safe way to use essential oils. Unfortunately, it eliminates almost all therapeutic benefit, and those trained in the British school often argue that essential oils provide no healing benefit, because they have not seen it. They believe essential oils may be inhaled, as in the German school, but they prefer to use evaporative diffusers that release tiny amounts of oil over long periods of time. They also believe that the use of essential oils can lead to sensitization and while essential oils do provide a wonderful, feel-good, relaxing experience, they should not be used or administered except by someone with years of training and hands-on experience.

Here in America, the melting-pot of the world's cultures, we have applied our melting-pot mentality to aromatherapy and combined the best of all three systems to create the American school of aromatherapy. While you will meet staunch proponents of one school or another here in the America, we are fortunate to have had the opportunity to learn from many of the pioneers in the field, including Daniel Penoel, MD, of France, and Jean-Claude Lapraz, MD, of France, Robert Tisserand, of the UK, D. Gary Young of the US, and Jeanne Rose, of the US. Penoel, Lapraz (both students of Dr. Jean Valnet) and Young (a student of Penoel and LaPraz) are recognized experts in the field and have taught classes throughout the world. Through

their efforts and the efforts of many others, aromatherapy has become popular the world over.

Now, there are numerous schools and individuals teaching the lay person how to use essential oils. For those of you who are interested in learning more about essential oils and their many benefits that go beyond a safe method for mold remediation, we recommend the following programs:

> The Center for Aromatherapy Research and Education (CARE) programs, developed by David Stewart, PhD, offer continuing education credits for massage therapists. This is an excellent hands-on program for anyone with an interest in learning Raindrop Technique developed by D. Gary Young, Healing Oils of the Bible, and about the chemistry of essential oils.

> The Institute of Spiritual Healing and Aromatherapy (ISHA) programs, developed by Linda Smith, RN, MS, offer continuing education credit for massage therapists, nurses, ministers, and also for Registered Aromatherapists (RA). They have several different program tracks, one for ministers and one that is approved by the National Association of Holistic Aromatherapists (NAHA), approved as Level 1 and Level 2 for individuals wishing to sit for the Aromatherapy Registration Council's RA exam. This program offers training in the historical uses of essential oils, the biblical uses of essential oils, and in the use of essential oils for energy and emotional healing. There are also classes offered in the Raindrop Technique developed by D. Gary Young.

In Canada, Dr. Sabina DeVita's Institute of Energy Wellness Studies (IEWS) program is the first federally accredited program (i.e., by the Canadian government) teaching what they call aromatic sciences. This is a comprehensive program offering training in Chemistry, Raindrop technique, energy healing, chemistry of essential oils, and much more. Canadian citizens are eligible to apply for a fully funded grant to pay tuition costs for participation in the program.

Each of these programs provides excellent training with only minor overlap in the materials covered. Contact information for each is included in the Recommended Resources section of this book.

What Are MSDS Sheets? What Is Their Purpose? Are They Reliable?

Professional contractors doing mold remediation and cleanup after flooding often ask us for the MSDS sheets for the products used in the 10-Step Protocol. Because this comes up often, we decided it was important to include a little background so that the reader would have a better understanding of MSDS sheets and the information they provide.

About 100 years ago manufacturers of dangerous chemicals began to supply information about the hazards associated with the use and handling of their products, to help their customers avoid serious accidents. Around 1945, the US Department of Labor published documents under the title: "Controlling Chemical Hazards," and about the same time the Manufacturers Chemical Association began publishing their "Chemical Safety Data Sheets" or their "SD" series of publications. When the US Congress enacted the "Longshoremans and Harbor Workers

Act of 1958," most the elements found today in MSDS sheets had been developed. However, it was not until November, 1983, that the federal Occupational Safety and Health Administration (OSHA) issued their final regulations for MSDS sheets in the Federal Register. Under this ruling, MSDS sheets were "required for all shipments of hazardous chemicals leaving the manufacturers work place and from all importers of the same on all shipments by November, 1985." Distributors and employers were required to comply with the new guidelines as of that same date. The general requirements are spelled out in the Code of Federal Regulations (CFRs), specifically 29 CFR 1910.1200: Toxic and Hazardous Substances, Subpart Z. (CFRs are now available on the internet).

OSHA requires MSDS sheets only for materials that 1) meet OSHA's definition of hazardous, and 2) are "known to be present in the workplace in such a manner that employees may be exposed under normal conditions of use or in a foreseeable emergency."

The regulations and the MSDS forms were put together from existing formats published by various chemical companies, State regulations, and associations. The sheet was tailored to meet the needs of a specific area of industrial uses, in particular longshoremen handling large crates and containers filled with hazardous materials.

MSDS sheets are not intended for use by the general consumer. They are intended for people working with hazardous materials in an occupational setting. For example, an MSDS for a cleaning solution is not highly pertinent to someone who uses the cleaner for short periods of time every day, or only occasionally, but can be extremely useful for someone who is exposed to the cleaner or its chemical constituents in a confined space for forty hours a week. MSDS sheets, designed to provide workers and emergency personnel with procedures for handling hazardous substances safely, include physical data such as melting point, boiling point, flash point, toxicity, lethal dose (LD50), health effects, first aid, reactivity, storage, disposal, protective equipment, and spill handling procedures.

The term LD50 (usually shown as LD_{50}) represents the dosage of a substance that results in killing 50% of the rats, mice, rabbits, or other types of animals used in a scientific study to determine the lethal dosage of a given substance. Every substance has an LD50, but it is not clear whether that information is of any great value, except in extreme circumstances.

Paracelsus, a Swiss physician and alchemist of the sixteenth century (1493-1541) said, "All things are poison and there is nothing that cannot be poison. The dose alone makes it poisonous."

Dr. David Stewart points out in his book, *The Chemistry of Essential Oils Made Simple* (CARE Publications, 2005), that numerous elements required for normal body function can be toxic, even fatal, in large doses.

MSDS sheets for essential oils utilize toxicological research prepared primarily for the food and fragrance industry. Dr. Jean Valnet has said, "...the body will become habituated to everything which is in any way adulterated, harmful or toxic..." and certainly most perfume-grade and food-grade essential oils are adulterated with potentially harmful and sometimes with toxic synthetic substances. That is why it is so important that you, the consumer, always use therapeutic grade essential oils from a trustworthy source.

There is great controversy about whether the toxicity for rats, mice, rabbits, etc., has any relationship to toxicity for humans. This topic continues to stir heated debate in the field of aromatherapy.

Robert Tisserand and Tony Balacs, in their book *Essential Oil Safety* (Churchill Livingston, 1999), say it is "largely an extrapolation of toxicological reports from the Research Institute for Fragrance Materials (RIFM)." This means the oils that were tested often were adulterated or standardized with synthetics that are known to cause allergic and toxic reactions in humans. The authors caution readers against attaching too much importance to acute oral LD50 figures, and in particular to acute dermal LD50 numbers, as "we have very little idea how well the animal data would extrapolate to the human situation."

Tisserand and Balacs report that LD50 values vary, even for the same substances, depending on the type(s) of animal(s) used, different laboratories, and differing methods of administration of the essential oil. So, there is good reason not to be overly concerned and not to place much importance on the LD50 for a given essential oil.

While we do not agree with many of the statements contained in Tisserand and Balacs' book, it does have value for those people who are specifically interested in learning about the LD50 for a given essential oil. Tables and text regarding the LD50 for essential oils are provided in grams per kilogram (g/kg). For most Americans this is a confusing way to think about how that information relates to uses by humans. For this and other reasons, we do not recommend this book for the general public.

For example, in one place in the book it states:

The average human adult weighs around 65 kg, so a lethal dose of cyanide would be 0.005 g/kg x 65 kg, or 0.325 g. One of the least toxic essential oils, valerian, has a toxicity level of 15 g/kg. For a 65-kg person, this is equivalent to a lethal dose of 975 g., almost 1 kilo of essential oil. This is around 2000 times the amount normally used in aromatherapy massage (15 ml of oil with 3% of essential oil).

To give you perspective, one kilo of essential oil is more than 35 ounces, or approximately 1,035 milliliters.

[**Note**: A gram is roughly equivalent to 1 ml of essential oil. A kilogram, or kilo, is approximately 2.205 pounds. Therefore, an LD50 of 15 g/kg for the essential oil of Valerian would mean a rat must consume roughly 15 ml of essential oil per 2.2 pounds of body weight for the essential oil dosage to be lethal.]

The average consumer uses essential oils in terms of drops, not milliliters. The maximum amount of essential oil being diffused during the 10-Step Protocol (see Part Four of this book) is 15 ml per 1,000 sq ft of space over a period of about one and a half to two days. It is **strongly recommended** that people and pets not be present in a room where intensive diffusing is taking place. This allows for maximum penetration and effectiveness of the essential oils, without excessive exposure to individuals or pets.

MSDS sheets for the Thieves® Household Cleaner that contains EOB2 (the Thieves® oil blend), and MSDS sheets for the individual essential oils in EOB2 (the Thieves® oil blend) are included in Appendix D. MSDS sheets are only for the individual essential oils, not for blends.

The MSDS sheets are included for those people who will be using these products in an occupational setting, where they may be required to provide MSDS sheets to OSHA inspectors.

The MSDS sheets were obtained from Young Living Essential Oils, the producer and/or distributor of the products used in the case studies presented in this book. They have been re-created, verbatim, for this book. You will notice that the format varies from sheet to sheet. This is because MSDS sheets for commercial products are prepared by each manufacturer, and the exact format of an MSDS can vary from source to source. Because we received many requests to do so, and because OSHA, cleaning services, and mold remediation companies require them, we have provided these facsimiles. As they have been reproduced verbatim, there are British spellings, misspellings, and a lack of consistency in the information provided. The authors are not responsible for the accuracy or consistency of the information contained in these MSDS sheets.

Please be advised that the authors disagree with some of the information contained in the MSDS sheets. In particular, the first-aid advice given in the MSDS sheets regarding contact with skin or eyes is, in our opinion, seriously in error. The direction to "flush eyes with large quantities of water" is a general admonition used in hazardous chemical MSDS sheets, and might be appropriate for some hazardous liquids. It is not, however, the best advice for dealing with exposure to essential oils.

Essential oils are not water soluble. The addition of water may spread the oil, and may cause the burning or irritation to intensify. Oils and water do not mix. **Essential oils are lipid soluble, meaning they will mix with other oils and are easily diluted when combined with fatty acid oils, such as cooking oils.**

This points to yet another problem with the reliability of the information contained in MSDS sheets. And, **for the average consumer, the information in an MSDS sheet is of little or no benefit.**

First-hand experience has taught us that when an essential oil or blend is inadvertently introduced into the eyes, adding a few drops of vegetable oil affords much quicker relief than flushing with water. If vegetable oil is not immediately available, then milk, honey, yogurt, butter, massage oil, vitamin E oil, or mineral oil will work far better than water.

The MSDS sheets (see Appendix D) for Cinnamon Bark (*Cinnamomum zeylanicum*), and Rosemary (*Rosmarinus officinalis*) are the only ones that list lethal dose information. Cinnamon Bark essential oil has an LD50 of 3.4 milliliters (ml) per 1 kilogram. LD50 information for Rosemary is given for both ingestion and skin contact when administered to rats and rabbits. The LD50 for ingestion of Rosemary essential oil for rats is greater than 5 ml/kg. A kilogram of weight is approximately equivalent to 2.205 pounds. The liquid measure 5 ml is roughly equivalent to about 1 teaspoon of liquid.

Converting these potentially lethal doses to an average adult weighing 150 pounds, we find that a person would have to ingest more than 340 ml (or approximately twenty-three 15 ml bottles) of Rosemary essential oil, within a short period of time. That is highly unlikely. The MSDS says that Rosemary's LD50 for skin contact for rabbits is greater than 10 ml per kilogram, or 680 ml, double that required for ingestion. The LD50 for Cinnamon Bark is 3.4 ml per kilogram, and for an average adult weighing 150 pounds that means they would have to consume more than 230 ml of Cinnamon Bark essential oil in a short period of time. Again, this is highly unlikely.

Taking this a step further, because LD50 ratings are based on using synthetic and perfume grade essential oils for testing on animals, this information may have absolutely no relationship to a given individual's response to an essential oil. Further, these oils comprise a small portion of the EOB2 blend, called Thieves® by the manufacturer, and LD50 for an individual essential oil does not tell us the LD50 for a blend, because it does not account for any buffering or quenching effects that may result from the individual oils being combined.

Are Essential Oils Non-Toxic?

As mentioned earlier, the LD50 varies based on the essential oil involved and the weight of the person (for additional information, see the book *Essential Oil Safety* by Tisserand and Balacs). And one should always bear in mind that the LD50 can be quite misleading and, in and of itself, is not a key to toxicity. However, looking at the LD50 acute oral toxicity information available for the individual essential oils contained in EOB2 (Thieves® essential oil blend) we find:

Essential oil	LD50
Cinnamon Bark	3.4 ml/kg
Clove bud	2-5 ml/kg
Lemon	more than 5 ml/kg
Eucalyptus radiata	no data available
Rosemary	5 ml/kg

According to Tisserand and Balacs, Eucalyptus (Taget) has an oral toxicity of between 2 g/kg and 5 g/kg. They also show *Eucalyptus citriodora* as having an LD50 that is over 5 g/kg). While information is not available for *Eucalyptus radiata*, it would be reasonable to use the lower toxicity rate for Eucalyptus because this means it takes less of the substance to effect death in rats. Therefore, to err on the side of safety, you can use the lower LD50. Tisserand and Balacs also tell us that essential oils with a lethal dose greater than 1.0 g/kg can be regarded as nontoxic.

Now, how does that relate to a human being weighing 150 pounds? How can you determine the LD50 for a child, a pet, or another human whose weight is different? Again, always remember that an individual's response may have nothing to do with the LD50. This is only provided for those who use the oils in an industrial setting.

First, you must convert American pounds to kilos.

150 pounds ÷ 2.205 = 68.0272 kg

Next use that number (pounds converted to kilos) and multiply by the number of grams or milliliters (ml) per kilo. For Cinnamon Bark with an LD50 of 3.4 ml/kg:

68.0272 x 3.4 = 231.2924 ml

An individual weighing 150 pounds would have to ingest 231.3 ml, or more than forty-six 5 ml bottles of Cinnamon Bark within a short period of time to reach the estimated lethal dosage for rats or mice. Clove oil has a range of 2 ml/kg to 5 ml/kg, so it requires roughly the same amount of Clove oil to reach the estimated lethal dose for rats and mice. Lemon, Rosemary, and *Eucalyptus radiata*, the essential oils in EOB2 (the Thieves® oil blend) would require at least as much, or more.

Based on the many references cited, the information available, and the levels of anticipated exposure for most end users, we feel it is safe to say that when EOB2 is used in accordance with the 10-Step Protocol, developed and defined in Part Four of this book, that EOB2 (Thieves® essential oil blend) and the EOB2 (Thieves®) Household Cleaner are non-toxic.

Dr. Jean Valnet writes, "Where aromatic essences are used, healing takes place quickly without dangerous toxicity." This is a very important statement.

Dr. Valnet also reported in his book, *The Practice of Aromatherapy*:

> Essential oils are especially valuable as antiseptics because their aggression towards microbial germs is matched by their total *harmlessness to tissue* – one of the chief defects of chemical antiseptics is that they are likely to be as harmful to the cells of the organism as to the cause of the disease...It is important to remember that antiseptics will destroy not only the micro-organisms but also the surrounding cells.
>
> Infectious germs do not appear to become accustomed to the essential oils as they do to the many forms of treatment using antibiotics. This is very important.
>
> [The body] cannot become habituated to the oils any more than it can become habituated to...olive oil...pure mountain air or personal hygiene. The results remain the same; they do not lessen over any length of time.

What does all this mean?

These statements suggest that micro-organisms (possibly including mold) are not able to adapt to essential oils, so essential oils are not likely to become obsolete or ineffective so long as they remain pure, and unadulterated. Essential oils are harmless to tissue, according to Valnet, and they are not habit-forming. These statements are very important and have been validated through the research of others, as well as through the use of essential oils by millions of people the world over. It may be that therapeutic grade essential oils offer us the safest substance that can be used for eliminating mold.

Based on the body of research that exists, it is safe to say that essential oils are, generally, non-toxic. However, like anything and everything else, they may be lethal when consumed in extremely large quantities.

It would be difficult, perhaps impossible, to inhale a lethal dosage of an essential oil, or to absorb a large enough dose through the skin. And although certain oils are of more concern than others, according to many researchers and writers, it is only through oral consumption, or oral overdoses that serious consequences have resulted. And even in the few rare cases where oral overdoses have occurred, it was either deliberate, or the accidental overdose when a child consumed bottles of essential oil. These are not normal, nor are they frequent occurrences, and they do not warrant the fears and concerns put forth in some books on aromatherapy. By contrast, pharmaceutical drugs result in large numbers of deaths from overdoses every year. Some estimates suggest that overdoses on properly prescribed medications exceed 100,000 deaths annually.

Deaths that may be attributed to an essential oils overdose are, by comparison, extremely rare. And the volume of essential oil that must be consumed to be lethal, while it varies from oil to oil and person to person, is so large for the essential oils used in EOB2 that it is hard to imagine that anyone using the oils in accordance with the 10-Step Protocol, presented in Part Four of this text, could overdose on the essential oils during a mold remediation treatment.

Having said that, it is important to remember that just as each of us has a different body

type, different blood type, different blood and body chemistries, individual responses to essential oils are likely to vary and for this reason it is always wise to follow certain safety precautions. And always closely monitor the responses of individuals and pets to the oils when diffusing. Some people and pets absolutely love essential oils, some do not.

Safety Tips and Precautions

1. Store essential oils in a cool, dry place, away from sunlight.

2. Do not drink an entire bottle of essential oil.

3. Do not put essential oils in the eye(s).

4. Do not put essential oils in the ears.

5. Only use essential oils in a diluted form in the bath.

6. One drop of an essential oil may be sufficient to effect a change. Using large quantities of essential oils on or in the body is almost certainly a waste of oil and money. More is not necessarily better when it comes to essential oils.

7. If you accidentally get essential oils in the eyes, it will burn. Apply liberal amounts of fatty acid oil to dilute the oil and burning will cease. We have found that applying water will only exacerbate the problem because essential oils are not water soluble. To dilute essential oils, reduce their action, and eliminate irritation to skin, eyes, ears, or if a large amount is accidentally consumed, we suggest you apply or consume extra virgin olive oil, vegetable oil, milk, butter, yoghurt, massage oil, vitamin E oil, or mineral oil until irritation subsides.

Essential oils are lipid soluble, meaning they are diluted by fatty acid oils such as vegetable and cooking oils.

Just so you know this suggestion really works, we will relate an experience we had with a client.

Margie decided to do an intensive diffusing (eight hours, non-stop) with EOB2 (Thieves® essential oil) as a follow-up to the mold remediation treatment completed in her home a month or so earlier. That evening, she called us in a panic. Her husband had set the diffuser on top of the kitchen table, right next to their family dog's bed, with the dog asleep in the doggie bed, and the diffuser had been running all day while they had gone to a nearby town to do shopping and other errands. When they returned, they found their beloved pet with its eyes rolled back in its head, and tongue hanging out. Margie was, understandably, in a panic.

We asked if she had any whole milk or yoghurt in the house that she could give to her dog, but she did not. We quickly ended the phone conversation so Margie could drive to the nearest grocery store.

When we did not hear back from Margie for a couple of hours, we gave her a follow up call. She told us that her dog had lapped up the yoghurt and the milk with great gusto, and then she was "Just fine."

Margie was so amazed at how well the suggestion had worked and how quickly her dog, Shiloh, had recovered that she saw no reason to call us back to let us know how things had turned out. She assumed we knew exactly what would happen.

Always follow up with clients. It would have been inappropriate, and we would have lax in our duty to our client, if we had not followed up and verified that Shiloh was well and that everything had returned to normal.

This story really drives home the need to be aware of safety precautions, and to have something on-hand to help if an emergency arrives. You can use essential oils and the 10-Step Protocol without fear, so long as you take into consideration the inhabitants of a space. Margie had been advised not to leave the dog in the room with the diffuser running whenever she was diffusing for more than fifteen minutes at a time. Her husband had not been present at the time that was mentioned. It is very easy to see how this situation might occur, so it is always good to be prepared and knowledgeable.

We have an expectation that essential oils will always work without doing harm, because we have seen it work for years. The oils have saved us from many doctor and hospital visits, as well as saving our clients from the high cost of standard mold-remediation treatments. However, we have also made efforts to educate ourselves about the safe use of essential oils and, in purchasing this book, you have done the same.

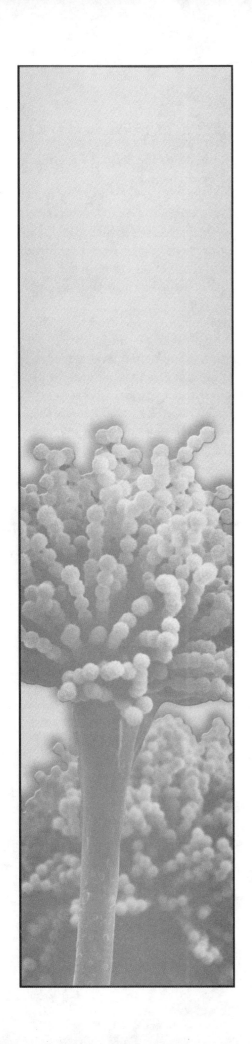

PART FOUR

THE CLOSE PROTOCOL

10 STEPS TO PREVENT AND ELIMINATE MOLD

We began this work from scratch. There were no protocols, no suggestions, no body of work or knowledge that told us how much essential oil had to be diffused or how long to diffuse the oils for them to be effective against mold. No documents existed that specifically said oils could be used to eliminate mold in buildings. We could not identify anything other than scientific papers, a few books with minor notations, and personal anecdotes that provided information about the antifungal properties of essential oils. There was no single book or document we could go to for guidance. And even with the information we had, it was not clear exactly which essential oils would work best in treating mold-infested buildings.

We utilized knowledge gained from the sources we had and from our personal backgrounds and interests, including experience in science, engineering, mold sampling and remediation, and aromatherapy. And, as is our custom, we also prayed and asked for God to guide us in developing experiments that would be true to good science and in developing a protocol that would provide value and worth to our clients, so that whatever happened, this work would serve a greater and, we hope, a higher purpose.

Diffusing EOB2 at different sites with different levels of mold infestation led to the development over time of an effective protocol for dealing with mold. In this chapter, we present the culmination of our efforts to date as a generalized protocol for mold prevention and elimination.

Our first attempt at this was published as the "10-Step Protocol to Effectively Control Toxic Mold" with the release in 2006 of the DVD, *Toxic Mold – A Breakthrough Discovery, Saving Health and Home.* That protocol was based on the case studies completed at that time. As more case studies have been completed, the protocol has evolved. Below is the latest, simplified version of what has become known as The Close Protocol. Following that are relevant details about each step in the process developed by the authors.

The Simplified Version of The Close Protocol (Patent Pending)

A. If you do not have mold, prevent mold by diffusing EOB2 for fifteen minutes every three to four hours at least once a week in each room.

B. If you have mold:
   1.   Sample.
   2.   Diffuse EOB2.

3. Repair leaks.
4. Clean thoroughly using the household cleaner that contains EOB2.
5. If sampling identifies toxic mold, and you have not already done so, begin intensive, non-stop diffusion of EOB2 <u>immediately</u> and continue diffusing non-stop for twenty-four to seventy-two hours in the spaces where the mold was found. Contact professional mold remediation companies to get estimates for having them remove and properly dispose of water-damaged and mold-infested materials.
6. Remove, repair, and replace water-damaged and mold-infested materials.
7. Re-sample.
8. Repeat Steps 1 through 7, if necessary.
9. Refinish affected areas.
10. Diffuse regularly to prevent mold (see item A above).

Preventing Mold Growth

Prevention is absolutely the best option. It will save you the most money. The cost of a diffuser and the essential oils blend is an investment in the health of you, your family, your pets, and/or employees, as well as in your property.

If you do not have a mold problem and no mysterious or chronic illness, then prevent mold growth from becoming a problem by diffusing EOB2 for fifteen to thirty minutes every three to four hours at least once each week in each room, or do a short, intensified diffusion of EOB2 once a week for eight hours in each room. Do monthly maintenance checks. Immediately repair any leaks and eliminate excess moisture wherever possible.

The old, familiar saying: "An ounce of prevention is worth a pound of cure," can be paraphrased with: Diffusing a few milliliters of EOB2 every week, can save you thousands of dollars.

Our personal experience is that diffusing EOB2 in our living space and offices has not only prevented the growth of mold and the potential costs associated with mold remediation, but has also saved us the misery of annual bouts with the flu, colds, bronchitis, sinusitis, and chronic allergies, almost entirely for the past eight years. We suffered with all of these before we started diffusing this essential oil blend on regular basis. We have also saved the expense of many doctor bills and pharmaceutical drugs, as well as the cost of over-the-counter medications and remedies. Better than that, we have enjoyed a higher quality of life. Life is more fun and far more productive when you are healthy. Thanks to the benefits we have personally received, thanks to the proof offered by the case studies in this book that this blend of oils is exceptionally good at eliminating molds, and thanks to the scientific research that has been done by others, we believe it is fair to say that diffusing EOB2 on a regular basis is good advice for everyone.

We have found that one diffuser is sufficient for about 1,000 square feet of floor space. Where we live and work, we have put diffusers on automatic timers. We add oil to the diffusers' wells every morning, and diffuse EOB2 for fifteen to thirty minutes every three to four hours. That is what we find works the best. We have a diffuser in our kitchen, one in each bathroom, one in each office and two in our basement, including one that we placed inside our heating/air conditioning unit.

You can always start with just one diffuser and move it from room to room. We use about one bottle of EOB2 (Thieves®) for every two diffusers per month, unless we have reason to do an intensive diffusing when more essential oil is required.

We also enjoy diffusing Lemon essential oil, a blend of citrus oils that includes Orange, Tangerine, Lemon, Mandarin, Grapefruit and a touch of Spearmint, as well as a variety of other individual or single essential oils and oil blends. We diffuse the EOB2 (Thieves®) blend every week. In general, if you are diffusing to prevent mold in a home or office that is 1,500 square feet or smaller, one 15 ml bottle of EOB2 will last an entire month. Just fill the well of the diffuser, put it on a timer, and check it every morning to see if oil should be added. It is really simple.

You can easily tell when one of the diffusers is running low on essential oil, because it sounds different. It will gurgle, sputter, and spit, making a louder noise when it is time to add more essential oil.

Most people find the aroma of EOB2 appealing. The Lemon, Cinnamon and Cloves in this blend remind us of the clean, fresh aromas of our grandmothers' kitchens. But everyone is different. And, if you are sensitive to the micro-mist of EOB2, diffuse the blend of oils when you are not in the space or not in the building. Diffuse for long periods and intensively when you are away, at work, on vacation, or out shopping. And diffuse for no more than fifteen minutes every four hours when you are present in the space. This will allow your body to become accustomed to the essential oils.

What Diffuser Is Best to Use?

A large variety of diffusers are available for dispersing essential oils. We have tried quite a number of them over the years and have found many disappointing. As mentioned earlier, the diffusers used in the tests described in this book were all purchased from a company in Utah, and we have been very happy with the performance of these diffusers during the eight years we have used them. They are sturdy, easily cleaned, and very effective at delivering a high volume of essential oil in a micro-fine mist that will permeate an entire room in a matter of minutes.

Numerous types of diffusers may be appropriate for many things, but our preference (especially when using the diffuser for the purpose of eliminating mold in buildings) is a cold-air diffuser that utilizes a non-reactive metal well to hold the essential oil, a glass nebulizer to disperse the oil in a micro-fine mist, and an air pump that forces the oil into and through the nebulizer. (The size of the pump used in most of the case studies was 115 volts, 60 hrz, 3.5 watts.) This type of diffuser will provide a concentrated micro-mist of essential oil at a high dispersion rate and will permeate the air in a space in just a matter of minutes.

Diffusers that use the glass nebulizer to hold the oil being dispersed have not proven effective in our experience. They are generally too small and not powerful enough to diffuse properly for this purpose of mold remediation. They also waste the essential oil, often requiring twice as much oil as the metal-well diffusers to operate properly. In our experience, these diffusers also tend to tip over and break more easily.

Diffusers that utilize a pad or wicking device of any kind are even less effective and more complicated. The pad or wicking device absorbs a small portion of the essential oil. There are problems diffusing some oils because they are too viscous (thick or dense).

We believe that all the compounds in an organically produced, therapeutic grade essential oil work together, in synergy, to provide benefits. If using a diffuser with a wicking device or a pad, certain compounds in the essential oils that are more readily absorbed by the pad or wicking material will be eliminated entirely, thus reducing the benefits and the results obtained. Also, these wicking devices do not allow the micro-mist of oil to be emitted into the air with the power or concentrated dispersion rates required to effectively permeate a space of 1,000 square feet. Therefore, we do not recommend using these diffusers for treating mold-infested buildings or for preventing mold, and personally we would not use for any purpose except in the situation where no other option is available, such as in a car.

Diffusers that utilize heat or water should be avoided. The diffusers that use heat destroy the most fragile organic compounds and also pose a fire hazard. Diffusers that use water and diffusers that use heat DO NOT disperse adequate amounts of the essential oil to be effective for mold prevention or remediation.

Diffusers that do not have an electric air pump to force cold air through the well of oil DO NOT disperse adequate amounts of the essential oil to be effective for mold prevention or remediation.

If You Have or Suspect Mold

Diffusing EOB2 will eliminate mold and keep it from coming back. However, it is very important to know whether you have TOXIC MOLD or not.

Step 1. Sampling

Toxic mold must be dealt with properly to avoid unnecessary exposure and potential health impacts. The only way to know if you have a toxic-mold problem is sampling, and when the subject of sampling comes up, the questions asked most often are:

Why? When? Who does it? and How?

Why Sample?

It is vital to remember that the most serious mistake anyone can make with mold is to underestimate the dangers of exposure to toxic mold. The health impacts have been explored in earlier chapters of this book. This point, however, cannot be overstated. Finding a *qualified* professional to take the appropriate samples and interpret the laboratory results is essential in order to determine what methods to employ in ridding a building of toxic mold (see the section "How to Find a Qualified Professional to Collect Samples" later in this chapter).

Sampling is often downplayed by professional cleaning and remediation companies. They will tell you that you have to tear out all visible mold- and water-damaged materials and get rid of it, so there is no need to sample. One remediation service's website boasts that their advice to avoid sampling will save you "loads of money." Their position is understandable; their focus is remediation. Plus, their efforts rarely solve the problem (see Case Study No. 15), so they are not altogether keen on having sampling done because it may, and quite probably will, show that all the money you spent to have them remediate was wasted. The truth is that proper sampling can actually save you thousands of dollars. Consumers will typically spend less than $100 on sampling for every $1,000 saved.

A List of Reasons Why You Should Sample for Mold

1. If you assume you have a benign mold and try to clean it up yourself, using rubber gloves, a sponge, and bleach water, you could be making a serious mistake. If, in fact, you have a toxic mold like

Stachybotrys chartarum, Chaetomium, or certain species of *Aspergillus or Cladosporium,* you could cause yourself and/or others serious health problems, and you could be spreading the mold rather than cleaning it up.

2. <u>Mold spores are not visible to the naked eye</u>. As we saw in several of the case studies in the Chapter 4, tens of thousands of mold spores per cubic meter can be present and not be seen, yet they are detected by appropriate air sampling (see Case Study No. 11 for an example). Without proper sampling, you really cannot know the nature and severity of the problem.

3. Remember, as we said in Chapter 2, all mold infestations are NOT alike, and even if you remove all visible molds, you may still have a hidden problem that could be identified with proper sampling.

4. A cleanup method that may be safe for some mold infestations can be disastrous for others.

5. Without adequate sampling, every remediation effort must be carried out assuming the mold you are dealing with is toxic, and that is a very costly approach.

When Should Samples Be Collected?

▪ Samples should be collected before diffusing the essential oil blend or attempting to remediate or clean up mold.

▪ Samples should be collected before purchasing a property.

▪ Collect samples any time you or someone in your home or office experiences chronic colds, flu, headaches, dizziness, nausea, nose bleeds, memory loss, or

other problems that disappear when away from the space for a few days or a week, and return shortly after returning to the space. Any or all of the symptoms noted above may be caused by toxic mold.

▪ When black mold is visible, sampling is a must. Black mold is not necessarily toxic, but some of the most dangerous mold species are black in appearance.

▪ Always sample as a follow-up to treatment or remediation to verify that the problem has been solved.

Can I Do the Sampling Myself?

No law or regulation says you cannot collect your own samples. However, in almost all cases, doing the sampling yourself will be a waste of your time and money.

When samples are collected by a qualified professional who has the proper training and experience, they know where to take the samples, what kind of samples to take, they have the proper equipment, and the results yield reliable information. The professional will also help explain the results and can assist you in locating reputable remediation contractors if you have toxic mold.

In later sections of this chapter we evaluate home test kits and mold test kits that are available at hardware and home improvement stores as well as over the Internet and explain why they rarely provide useful information.

How to Find a Qualified Professional to Collect Samples

Please be aware that most professionals do not yet know about the efficacy of using the essential oils blend tested in this book for treating mold. This option for ridding your life of toxic mold is NEW. Most professionals will have other recommendations, which you should consider, but there is no requirement that you follow their advice any more than the advice in this book. It is important, however, that you have a professional collect samples for you. So, how do you find a qualified professional to do mold sampling?

1. Contact the Indoor Air Quality Association (IAQA) and ask for referrals to qualified professionals in your area. The association's website is www.iaqa.org, which has a searchable list of members.

2. Locate an environmental consulting firm that does mold sampling or that specializes in indoor air quality. Often this is as simple as looking in the yellow pages of your local directory for environmental consultant, environmental engineering, industrial hygiene, or mold sampling.

3. Contact a professional industrial hygienist. The American Industrial Hygienist Association has a searchable list of consultants and professionals, many of whom are qualified to do mold sampling. Visit their website at www.aiha.org.

4. Contact a university in your area that has an environmental science or an engineering program. Ask for a recommendation about a qualified professional in your area who does mold sampling.

5. Go to your state's directory of licensed professional engineers, and locate engineers in your area. Contact one of them, and ask if he or she does mold sampling, or can refer you to a qualified environmental engineer or another professional who does.

Requirements for certification and registration of professionals vary from state to state, and there are currently no federal standards established for levels of mold that constitute a health hazard. However, someone who has sufficient experience in mold sampling and remediation will have a good background on which to base professional opinions and judgments.

Questions to Ask a Qualified Professional

Once a qualified professional has been located, then be sure to ask that professional the following questions:

- How long have you been doing mold sampling?

- What type of training, credentials, and background in mold sampling and remediation do you have?

- What type of samples will you collect?

- Do you provide a report, a copy of the sample analyses, and an explanation of your findings?

After sampling is completed, and the professional contacts you with the analytical results for the samples collected, he or she will explain the results and offer suggestions as to how you should proceed. Often they will have their own favorite remedy which, based on our experience, will almost certainly not be as safe or as effective as using EOB2 and the 10-Steps Protocol. This has been proven in the tests performed and documented in this book. Also, other methods are usually far more costly than using essential oils. So, when you have located someone who has the credentials, experience, and equipment to do a proper job of sampling and analysis, you can avoid overreaction and added expense by doing the following:

1. Make sure the sampler takes both indoor and outdoor spore-trap air samples, so you will know what is growing inside the building. Tape lift or viable bulk-samples may also be necessary.

2. Start diffusing the essential oil blend immediately AFTER the samples are collected. Do not diffuse before the samples are collected because this will affect the results of the sampling.

3. Ask for a written report identifying the mold species found inside the building.

4. If the sampling company does remediation or recommends someone, thank them politely, and tell them you will definitely consider a proposal from them.

5. If there is no toxic mold present, and mold spore counts are not extraordinarily high, then you can diffuse the essential oil blend intensively for at least 24 hours in the rooms where mold exists. Afterward, use the household cleaner containing EOB2 to clean visibly infested areas.

6. **If toxic mold is found**, then be certain to diffuse the essential oil blend for a minimum of 24-72 hours in the areas where the mold was found.

7. After the intensive diffusing, contact your professional again, say you have used a non-toxic fungicidal agent to eliminate the mold, and that you would like re-sampling done to determine whether any toxic mold is still present in the air.

8. Discuss the results with your professional. Work closely with him or her to develop a plan for appropriate cleanup and removal of any remaining mold infestation.

   You have the right to demand that they use the non-toxic agents of your choice to do any cleaning of areas that have visible mold.

9. **If you have toxic mold**, you will need a professional's help to identify the source(s) of the infestation, the areas where leaks or moisture are creating the conditions necessary for mold growth, and in removing and disposing of toxic mold-contaminated materials.

EOB2 (i.e., the Thieves® essential oil blend) and the Thieves® Household Cleaner are the best treatment options available today. Not only are they more effective than other treatments, they are the only options that are non-toxic, can be used long-term to prevent the recurrence of mold, and have been proven by scientific study to destroy bacteria and viruses. The Thieves® blend is also approved by the FDA as a food supplement. No other mold treatment option can say that.

Always remember:

- If you have a professional remediation contractor or a cleaning service do the mold cleanup and remediation, do not accept the use of bleach, chlorinated petrochemicals, or any other toxic chemicals. You have rights.

- If you do any of the cleanup yourself, be sure to obtain the proper protective equipment so that your health and vitality are protected.

Over-the-Counter Sampling Kits

Over-the-counter (OTC) sampling kits are, at first glance, cheap, and they come with instructions for the property owner concerned about a possible mold problem. However, the information that a prepackaged kit can provide is extremely limited, which in the long run can make it far more costly. For example, sample results from some over-the-counter kits only tell you whether mold spores are present, not the concentration or species. This information is nearly useless, since mold spores are present virtually everywhere.

> **The most serious mistake made by property owners with mold problems occurs when toxic mold is treated with bleach or other toxic chemicals and mold-infested materials are handled without using the proper protective clothing and equipment.**

The kits that do identify mold species typically depend on culturing mold spores or structures from settling plates (Petri dishes) or sterile cotton swabs that are to be rubbed across a surface or twirled in the air to collect mold spores. Because the surface area or volume of air sampled by such kits is very limited, and mold growth and spore concentrations generally vary from room to room in any building, the results will usually be misleading.

To put it simply, with an OTC sampling kit, you are only going to know what was in the immediately surrounding area where the sample was collected. To get representative samples, you would have to have one sample collection of this type for about every five square feet of space in a room. At that rate, you can hire a professional to collect reliable samples and will also receive the added benefit of his interpretation of the sampling results for about the same cost.

Field comparisons have shown that these types of mold-sampling kits often fail to detect serious mold-infestation problems that are quickly identified by an experienced professional. Using these over-the-counter test kits could result in someone suffering

chronic or long-term illness as a result of the real source of the problem going undetected.

Over-the-counter test kits may yield some limited information, but the user must understand exactly how to use the kit, what information the kit is capable of providing, and how to interpret the results properly. Negative test results from one or two swabs or collection plates, for example, should never be relied upon to conclude that a building is free of mold-related problems. And positive test results from kit samples should only be considered a starting point. With heavy mold infestations, especially if toxic mold is present, having the sampling and interpretation of the results done by a professional is highly recommended.

Professional, third-party sampling is generally required by lending institutions and insurance companies. This is especially true in legal proceedings. Even when legal proceedings are not involved, and you just want to get rid of a mold infestation, you should have sampling done by a professional to make sure you know what you are dealing with and what your options are.

Do-it-yourself sampling can also be dangerous. The most serious mistake made by property owners with mold problems occurs when toxic mold is treated with bleach or other toxic chemicals and mold-infested materials are handled without using the proper protective clothing and equipment.

The least serious consequences include short-term illness with irritation to eyes, ears, skin and the respiratory system. These symptoms can be caused by breathing the toxic fumes of the bleach or other chemical fungicide mixed with toxic-mold spores. In some well-documented cases, exposure to toxic mold has resulted in hospitalization and/or long-term health problems.

Economic impact, even though arguably less important than health, can also be substantial. While inspection and sampling by a qualified professional may cost anywhere from a few hundred to a few thousand dollars (depending on the size of the building, the number of rooms, and the geographic location of the site), it can save several thousands of dollars in lost time, medical bills, and additional remediation expense.

Sampling Methods, Their Advantages and Disadvantages

Several methods are utilized in mold sampling, including the five listed and discussed below. One or more of these methods will be appropriate to different site-specific circumstances. New sampling methods are being devised every year for specific conditions and needs. For example, if a high level of certainty is necessary for legal proceedings, species identification using DNA sequencing may be appropriate.

Proper sampling is not a simple matter of swabbing a moldy spot and sending it to a laboratory. To avoid costly mistakes and determine the appropriate procedure for dealing with a specific mold problem, the following questions must be answered:

1. How many and what types of mold species are involved?

2. How many mold spores are there of each species per volumetric unit of air?

3. Where are the mold colonies located?

4. What are the sources of the mold spores, moisture, and mold food?

5. How long has the mold problem existed?

6. What is the extent of the infestation?

7. What are the ages and states of health of the persons exposed?

These questions can be only answered definitively by a thorough inspection using appropriate sampling procedures. The five sampling methodologies below are the sampling techniques used in the case studies described in Chapter 4.

> Impactor Air Sampling
> Viable Air Samples
> Tape Lift Samples
> Bulk Samples
> Wall Probes

Impactor Air Samples

Laboratory analyses of impactor or spore-trap samples will determine items 1 and 2, above. Outdoor and indoor samples MUST be taken in order to determine which mold species are growing inside. Outdoor samples should be collected at least fifty feet from the building.

This method consists of setting up an air pump and pulling air though a specially manufactured device that traps the spores that happen to be in the air. The pump must be accurately calibrated so that you know exactly how much air it moves per unit of time in order for the laboratory to provide accurate results. The time and pumping rate are recorded so that the exact volume of air pulled through the trap can be documented on the chain-of-custody (COC) form submitted to the laboratory with the sample. In the lab, a microbiologist determines the total number of spores of each of the mold species per cubic meter of air.

Advantages:

- If the pump is operated for a sufficient length of time and a sufficient volume of air is sampled, the results are representative of an extended space, such as a room or office.

- If the results are compared with a spore-trap sample of the ambient air outside, the species of mold suspected of growing inside the space can be identified.

Disadvantages:

- Results can vary due to any of a number of activities disturbing the mold and spores in the test space, and to a minor degree, the variability of laboratory analytical methods.

- Some mold species cannot be specifically identified using this method. For example, *Aspergillus* and *Penicillium* spores are indistinguishable in this method and are identified in the count provided on the lab's report as *Penicillium/Aspergillus*.

- This type of sample provides information on spore counts but does not distinguish between dead and living (viable) spores.

Viable Air Samples

An air pump is used to collect spores on an agar plate in a special type of impactor. The agar provides a base or substrate on which the mold spores can grow.

Advantages:

- This type of air sample allows the lab to identify living (viable) mold spores that can produce new colonies of mold.

- The professional consultant is able to make a more reliable determination of which mold species are growing inside the building.

- Better identification of individual mold species is possible.

- Differentiation between *Aspergillus* and *Penicillium* is possible.

Disadvantages:

- It takes longer to get results.

- Viable sample analysis is usually more expensive than nonviable.

Tape Lift Samples

Direct examination by a microbiologist or mycologist will determine the specific mold species growing on the tape. In principle, this method is very simple: A piece of clear tape is placed on the mold, then lifted from the surface carefully and placed in a sealable clear plastic bag and shipped to a microbiology lab for analysis.

The advantage of this method is its simplicity.

The disadvantages are:
- It only yields information about the one small spot where the tape was pressed.

- The structure of the mold may be crushed or smeared, making identification of species difficult.

- The mold growth may be masked by lint, particulate matter, and other debris stuck to the tape.

Bulk Samples

A piece of material (about one inch square) cut from the surface upon which mold growth is visible is collected, sealed in a plastic bag, and shipped to the laboratory for analysis.

Advantages:

- Bulk samples are better than tape lift samples for positive identification of mold species because growth structure is not damaged.

- Like viable air samples, the sample can be cultured to determine which mold species are alive and growing on the material.

Disadvantages:

- Cutting damages the material, eliminating the possibility of cleaning it and leaving it in place.

- Like the tape-lift method, this type of sample only yields information about the one small spot from which it was collected.

Wall Probes

A small hole about one-fourth inch in diameter is drilled through the wall behind which a mold infestation is suspected. A probe is inserted with an airtight seal, and air is drawn from inside the wall cavity into a spore trap or impactor.

Advantages:

- A wall probe sample allows you to determine whether mold growth exists inside a wall cavity without the time and materials expenses involved in removing a section of the wall.

- This method avoids disturbing mold spores and is, therefore, safer than more invasive procedures.

- The hole can be easily repaired with spackling or putty and paint.

Disadvantages:

- It may be difficult to determine exactly where to drill.

- Many holes may have to be drilled in structures with solid studs and sills that create barriers to air movement within the wall.

Step 2. DIFFUSE: Eliminate Mold Spores from the Air by Diffusing EOB2

After samples have been collected, diffuse EOB2 for 24 to 72 hours non-stop in the space or spaces where mold is known or suspected to exist.

Experience has shown that one cold-air diffuser of the type used in the case studies is effective in areas of up to 1,000 square feet in floor space. For best results, close all doors and windows in the room, and leave them closed during diffusion to achieve maximum penetration and absorption of the micro-mist of EOB2.

If sampling results indicate toxic mold, avoid toxic chemicals, and eliminate the mold by intensive diffusion of EOB2 combined with spraying source areas (crawl spaces, plumbing, HVAC ducts, and other closed spaces) with EOB2 and/or the household cleaner containing EOB2. If cleaning porous materials or spraying large areas, be sure to use the household cleaner undiluted. Even small amounts of water can provide all the moisture necessary for some toxic-mold species to grow. In some cases, where toxic-mold levels are high, it may be advisable to add an additional 15 ml bottle of EOB2 to the household cleaner before spraying in an area. Based on the results obtained during the case studies documented Chapter 4, we have concluded that this is the safest, most effective at removing mold, and also the most cost-effective method of mold remediation currently available.

Step 3. REPAIR LEAKS: Eliminate Excess Moisture

Preventing mold infestations in existing buildings primarily involves controlling moisture on surfaces that can support mold growth. **The following actions are basic mold-prevention measures that can save your health and your home.**

1. Fix water leaks and prevent excess moisture.

2. Control indoor humidity.

3. Conduct regular cleaning of and inspection of areas where moisture may create mold problems.

4. Make sure you have proper ventilation.

5. Keep mold-infested materials out of the building

6. Reduce and/or eliminate mold spores in indoor air by diffusing EOB2 regularly.

Preventing Water Leaks and Excess Moisture

The most common reason that molds start growing inside a building, causing health problems for the inhabitants, is excess moisture. Excess moisture results from high levels of indoor humidity causing condensation on exposed metal surfaces, water leaks, and other forms of water intrusion into the home or building. Moisture from condensation can penetrate building materials such as sheetrock, wallpaper, and wood, and create an ideal environment for mold spores to start colonies. Covering metal surfaces with insulation helps to prevent condensation.

Eliminate dead-air spaces by ventilating crawl spaces and basement storage areas. Line crawl spaces with one or more layers of polyethylene to cover earth contact areas that may allow dampness to rise into the ground floor of the building. Make sure that floors and walls are properly sealed to prevent moisture entry.

Leaking Pipes

It is often difficult to detect small leaks from water pipes until it is too late to prevent mold growth, since water pipes are normally concealed from view. The following are signs that may indicate leaking pipes:

- Running water sounds when nothing is turned on.

- Musty odors under sinks or near walls or floors.

- Flowing water in toilets.

- Dripping faucets.

- Abnormal increases in water bills or water meter readings that increase without increases in water usage.

- Discolored or water damaged walls or ceilings.

- Dampness around cracks in basement walls or floors.

- Abnormally high humidity in one or more rooms.

- Warm spots on concrete slab floors.

- Mildew or excess moisture under carpets.

- Sewer backups.

- Lawn areas that are unnaturally wet, with atypical growth of plants or grasses.

Water Buildup around Building Foundations

Prevent water from rainfall or lawn irrigation from collecting around the foundation and penetrating walls or floors by making sure the ground slopes away from the building. It may be necessary to improve perimeter drainage by installing a subsurface drainage system of crushed stone and screened or perforated pipes.

Roof Leaks

A common building practice, especially for offices, schools and hospitals, is to install air conditioning or air-handling units on a flat roof. Because of vibrations, maintenance personnel walking on the roof and improper or incomplete sealing around rooftop structures, leaks develop.

Another common cause of roof leaks is cracks between roofing material and skylights, chimneys, vents, etc.. In the case of a very large chimney or fireplace, a metal or tile cap should be mounted on top to prevent precipitation from falling directly into the opening. Also look for water stains on the ceiling of rooms, on rafters, and in the attic. Wooden beams that have absorbed water support mold growth and become weakened.

Ice Damming

A way unwanted moisture may invade a building during freezing weather is by a phenomenon called ice damming. When ice or snow cover a roof that does not have proper insulation and ventilation, the heated air in overhead spaces will warm the underside of the roof decking and melt the bottom layer of ice. Ice dams are formed when this water runs down the roof until it reaches the gutter and freezes. With no escape route, water pools and backs up and then

enters the building through nail holes and seams in the roof decking.

Windows

Condensation on or around the window is a sign of inadequate insulation and/or excessive moisture. We have seen mold grow on window sills and even on glass panes and aluminum frames. This source of excess moisture can be eliminated by installing storm windows and/or using a sealant around the window to make sure that it is indeed airtight.

Controlling Humidity

It is important to keep relative-humidity levels in all parts of a building below **55%**. Toxic molds, bacteria, and dust mites thrive when the relative humidity is at or above this level. It is a good idea to monitor humidity so you know when it reaches the danger zone, and then you can dehumidify, when necessary. In order to know whether you have a humidity problem and which areas or rooms are at risk, it is necessary to monitor the humidity in each room because humidity levels, like mold-spore concentrations, may vary from room to room throughout a building.

If you suspect you have a humidity problem, install relative-humidity sensors, called hygrometers or moisture meters. Each room or enclosed space should be monitored separately, especially bathrooms, basements, crawl spaces, and cellars. It is also a good idea to monitor humidity levels in ductwork, especially near the air-handling equipment and filter locations and inside or between walls. It is not unusual to find the source of a mold problem in HVAC ductwork. See Case Study No. 11, in Chapter 4, for a good example.

Step 4. CLEAN THOROUGHLY

Use the household cleaner containing EOB2, undiluted to clean all areas of visible mold. If necessary, add an additional 15 milliliter bottle of the EOB2 blend to the household cleaner. The household cleaner with EOB2 may be sprayed in crawl spaces and on metal ductwork. It may also be used to clean tiles, sheetrock, and other porous materials that have small areas of visible mold.

Note: This step should _never_ be undertaken before Steps 1 & 2.

If the results of proper sampling, as described in Step 2 above, show no toxic-mold species, and you decide to do the cleaning yourself, or if you have the help of family members or other nonprofessionals, be sure that everyone involved has and uses adequate protective equipment. The primary function of personal protective equipment (PPE) is to avoid inhaling mold and mold spores and to avoid mold contact with the skin or eyes.

Adequate Personal Protective Equipment (PPE)

- When cleaning surface areas of visible mold, you should use at least an N-95 respirator covering the nose and mouth. An N-95 respirator, which will filter out 95% of the particulate matter in the air, can be purchased at a home supply store or a hardware store.

- For adequate eye protection, goggles that are designed to prevent the entry of dust and small particles are recommended. Safety glasses or goggles with open vent holes are not acceptable.

- Disposable clothing is recommended to prevent the transfer and spread of mold to clothing and to eliminate skin contact with

mold. Mold-impervious disposable head and foot coverings, and a body suit made of a material such as TYVEK® should be used and all gaps, such as those around ankles and wrists, should be sealed with duct tape.

- Gloves are required to protect the skin from contact with mold, allergens, mold toxins, and cleaning solutions. Long gloves that extend to the middle of the forearm are recommended. The glove material should be sturdy enough to withstand abrasion from the materials being handled.

- Ear plugs should also be used because hearing loss is often reported by those exposed to high concentrations of airborne mold spores.

Step 5. IF STEP 2 SAMPLING RESULTS IDENTIFY TOXIC MOLD

Begin intensive diffusion of EOB2 immediately and continue non-stop diffusing for 24 to 72 hours. Contact a professional mold remediation company for estimates and assistance in removing and properly disposing of water-damaged and mold-infested materials.

Note: Removal of mold-infested materials should _never_ be undertaken before Steps 1 & 2.

Step 6. REMOVE, REPAIR, and REPLACE Water-Damaged and Mold-Infested Materials.

If toxic mold is present, huge numbers of spores will become airborne as materials are hammered and pulled apart, creating a health hazard and the potential spreading of toxic mold.

If *anyone* in the building is not wearing an N-95 dust mask or respirator, he or she will inhale large numbers of mold spores. Serious exposure to the dangers of toxic mold often occurs when mold is discovered during demolition or renovation of a building. When mold is found behind wallpaper, in HVAC systems, or around plumbing while removing building materials, the temptation is to tear out the mold-infested materials and throw them into a dumpster. When this is done, thousands of spores become airborne, and a serious health hazard is created.

One example we have encountered is that of a healthy man in his late thirties who was renovating a bathroom in his father's house. He found black mold growing behind the shower enclosure when he removed it, and without the benefit of any sort of protection, he tore the infested materials out and piled them in the backyard. Within a few hours, he became very ill, vomiting and sweating profusely. Two weeks later, he was still experiencing headaches and dizziness. Only then did he realize that his problems were caused by exposure to toxic mold. He may suffer long-term health problems from this exposure, and this could have been prevented, if he had known what he was dealing with and taken proper precautions.

Removal of mold-infested materials involves tearing out moldy porous materials, such as wallpaper, sheetrock, and plywood. Such activities disturb mold, and mold spores become airborne, increasing the risk of exposure.

Note: Before you begin this phase of mold remediation, be aware that federal regulations require that all individuals using certain PPE equipment, including full-face respirators, must be trained, must have medical clearance, and must be fit-tested by a trained professional. In addition, the use of respirators must follow a complete respiratory protection program as specified by the Occupational Safety and Health Administration (OSHA). For these reasons, in addition to safeguarding your health and the health of your family or employees, a qualified professional mold remediation company should be contracted to remove toxic mold.

After sampling, treatment, and cleaning have been completed, then water-damaged and mold-infested materials may be removed.

Be certain that all materials are sealed in plastic bags before disposal to ensure the safety of others, including neighbors, as well as disposal and landfill personnel. Remember, many species of toxic-mold spores may be spread with relative ease through something as minor as air movement.

Step 7. RE-SAMPLE

Collecting samples again following all remediation efforts is absolutely necessary to be sure the problem has been solved. If the protocol has been followed, and mold spores are still present, especially at elevated levels, then the source or an additional source of the mold may not have been discovered.

If, after removing all visible water- and mold-damaged materials the mold-remediation contractor says, "You can't see or smell any mold, so you no longer have a mold problem—" BEWARE!

Remember, you cannot see mold spores. Even professionals can be misinformed. If re-sampling reveals one or more species with elevated spore concentrations, molds are still growing somewhere in the building.

Chapter 7

Step 8: REPEAT Steps 1 Through 7, If Necessary

Among all the opportunities we have had to solve toxic-mold problems with EOB2 so far, only one required repeating Steps 1 through 7. The reason however, was not because the first application had not worked; it was because a different source of mold spores, a previously undetected growth of *Stachybotrys chartarum* colonies, had not been discovered. This step is included in the protocol expressly because this could happen in any building where there are multiple locations with conditions conducive to mold growth.

Step 9: REFINISH Affected Areas

This is the final step in remediating mold. This step is necessary only in cases where mold-infested and/or water-damaged materials like wallpaper, sheetrock, wood paneling, walls, ceilings, wood framing or other structural materials have been removed. This step should not be undertaken until sampling results have verified a substantial or complete reduction in mold spores of concern.

Step 10: DIFFUSE REGULARLY. Prevent the Recurrence of Mold

We strongly recommend continuing to diffuse EOB2 fifteen minutes every three to four hours at least three times a week as a preventative measure. Conditions conducive to mold infestation can occur in any building wherever excess moisture appears. As a building ages small leaks may develop in roofs, around windows, in piping, or in other locations, and these may go unnoticed until it is too late to prevent mold growth.

The documented case studies in this book demonstrate that diffusing EOB2 will remove

mold spores and keep them out of the air, offering protection even against hidden mold growth.

How to Become a Qualified Professional Mold Inspector

Occasionally, seminar and program attendees ask us how they might become certified to serve as a mold inspector and collect samples for clients. An indoor air quality (IAQ) consultant is a professional who can identify the causes of poor indoor air quality even when the problems seem vague or unrelated to visible causes. Such a person should be trained to see a building as an organic whole, with dozens of interrelated systems contributing to the overall health of the space and the people who occupy it.

It is an honorable calling and one that has the potential to affect the lives of many in a positive and meaningful way. To become an IAQ consultant takes years of education and training. However, there are many opportunities for those who wish to be able to help others by doing mold inspections, sampling, and cleanup.

At the time of this book's writing, few laws and regulations govern the field of mold inspection or home inspection, and a large number of training programs are available to those interested in these fields. Laws vary from state to state, so always check with your local licensing bureau and with professionals who do this type of work in your area. Only one training program has achieved a third-party accreditation from the Council of Engineering and Scientific Specialty Boards, and that program is offered by the American Indoor Air Quality Council (AmIAQC).

The AmIAQC offers certification examinations to those who have a minimum of two years of field experience. This program is positioned to become accepted nationwide. You can visit the organization's website for more information on its certification exams at: www.iaqcouncil.org/Professionals/ certifications.htm.

The place to start, if you are interested in learning more about becoming a qualified, professional mold inspector, is the Indoor Air Quality Association (IAQA). This organization offers courses that provide training and prepares candidates for taking the AmIAQC's certification examinations. The courses are offered in various locations throughout the United States and may also be taken online. Visit them on the web at: www.iaqa.org, and click on "Education."

A Message of Hope

If you have chronic or sudden unexplained health problems, like so many of the people whose stories are in this book, and especially if those health problems are respiratory in nature, then we urge you to consider the possibility that mold may be a causative or contributing factor. If you have suffered for years with illnesses that only seem to get worse and never better (as some people reported to us and never thought for a moment it could be a result of mold exposure), then it might be worth the expense to sample and either confirm or eliminate the possibility that mold is contributing to your diminished quality of life.

We end with this message of hope for all who suffer from chronic allergies, sinusitis, ear infections, chronic colds and more.

You may be suffering, needlessly, as a result of exposure to mold spores. There is an answer. One that is non-toxic, that has been provided by Nature, and when used appropriately, it will open the door to a new, healthier life, a new beginning for each and every one of you.

We believe in Divine Guidance, and if you have found this book, or received it as a gift, we believe it is because the Divine is answering a prayer, or offering you an opportunity, or opening a new door to a greater understanding of the abundant blessings God has prepared for each and every one of us.

This new discovery that essential oils can eliminate mold and keep it from coming back is monumental. It offers a cost-effective means of remediating mold and at the very same time provides a long-term solution. No other product that eliminates mold can be used in the home or office on a daily basis without adverse health impacts. No other product that eliminates mold is sold as a food supplement.

The data have proven there is a non-toxic solution to toxic mold, and it has been right under our noses all the time.

The Close Protocol
To Prevent and Eliminate Mold
(Patent Pending)

A. If you do <u>not</u> have mold – **PREVENT MOLD** by diffusing EOB2 for 15 minutes every 3-4 hours 2 to 3 times each week, or by diffusing continuously for at least 8 hours once a week.

B. **If you <u>have</u> mold** – Use EOB2 to eliminate mold and keep it from coming back.

    1. **SAMPLING** – Before diffusing, have samples collected by a professional to determine whether you have toxic mold. Toxic mold must be dealt with differently than non-toxic mold. Do not diffuse EOB2 before samples are collected.

    2. **DIFFUSE** – Eliminate mold by diffusing EOB2 continuously for at least 24 hours, and using the appropriate protocol for either toxic or non-toxic mold infestations.

    3. **REPAIR LEAKS** – Eliminate all sources of excess moisture.

    4. **CLEAN THOROUGHLY** – If sampling identifies <u>**NO**</u> <u>toxic mold</u>, then after diffusing, clean visible mold and stains with the household cleaner containing EOB2. Protect yourself from dust and mold spores by using proper protective equipment.

    5. If sampling identifies <u>**TOXIC MOLD**</u>, and you have not already done so, begin intensive diffusion of EOB2 <u>immediately</u>. Contact professional mold remediation companies to get estimates for removing and properly disposing water-damaged and mold-infested materials.

    6. **REMOVE, REPAIR, AND REPLACE** – If sampling revealed no toxic mold, remove water- or mold-damaged materials, and seal them in plastic for disposal. Non-toxic materials may be placed in normal garbage disposal containers. If sampling identified toxic mold, be sure to diffuse continuously for a minimum of 24 hours after your professional remediation contractor has removed all mold-infested materials.

    7. **RE-SAMPLE** – Have a professional re-sample for an objective determination and documentation of the effectiveness of Steps 1 through 5.

    8. **REPEAT** – Repeat Steps 1 through 7, if necessary.

    9. **REFINISH** affected areas.

    10. **DIFFUSE REGULARLY** - to prevent recurrence of mold infestation.

PART FIVE

APPENDICES AND RESOURCES

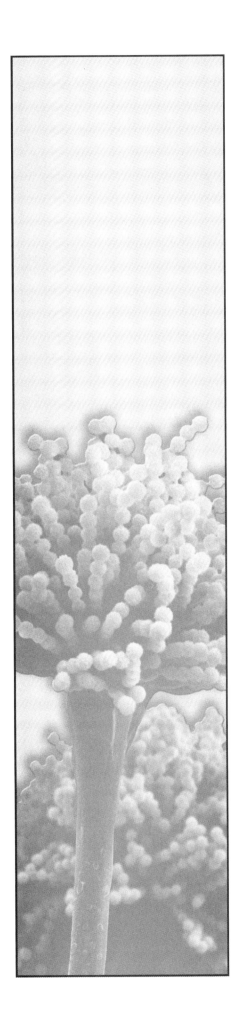

APPENDIX A

AN EDUCATIONAL PRESENTATION YOU CAN DO

An Educational Presentation You Can Do

We have been asked to provide a concise, educational presentation that would allow others to inform family members, friends, or a professional doing mold remediation about the facts and the discoveries made and presented in this book. Permission is granted to freely photocopy this chapter of this book for these purposes only. A beautiful, expanded, color version of

this information (which includes many photographs and Dr. Close's Patio Test) is also available for a fee as a PowerPoint presentation. It may be used on laptops, for slide shows, and to print out color pages to use in creating 3-ring binder presentations. To purchase the expanded color presentation, visit **www.MoldRx4u.com**.

TOXIC MOLD – A BREAKTHROUGH DISCOVERY

Disclaimer: This information is for educational purposes only. Individual results may vary. The authors and the publisher shall have neither liability nor responsibility to any person or entity for the use or misuse of this information. This information is not a substitute for medical counseling with a licensed health professional

This information is excerpted from the book, *Nature's Mold Rx – The Non-Toxic Solution to Toxic Mold,* 2007, by Edward R. Close, PhD and Jacquelyn A Close, RA. The Closes first began doing research into using essential oils for eliminating mold in November, 2005. Dr. Close has a PhD in environmental science and engineering, is a Registered Professional Engineer in the State of Missouri, and served as environmental advisor to more than fifteen Fortune 500 companies during his forty years experience in the industry. Jacqui Close is a Registered Aromatherapist with more than ten years experience with essential oils. The book includes twenty case studies and the following:

More than 50% of Homes Have Mold – Resulting in a 50 to 100% Increase in Respiratory Problems, according to studies performed by Harvard University and Johns Hopkins University.

It is estimated that 45 million Americans suffer from allergies. Over 50% of patients tested in one allergy clinic who had allergy symptoms tested positive for the mold species *Cladosporium.*

According to a Mayo Clinic study, 1 in 7 Americans suffers from acute fungal sinusitis.

A short list of the diseases that have been shown by scientific studies to be caused or aggravated by mold and, in particular, by toxic mold includes:

Headaches, Allergies, Colds, Sinusitis, Flu-like Symptoms, Memory Loss, Chronic Fatigue, Dermatitis, Fibromyalgia, Sudden Infant Death Syndrome, Bleeding Lungs, Brain Hemorrhaging, Nose Bleeds, Lung Disease, Asthma, and many more. You can also die from exposure to mold.

Using Bleach, Ozone, Toxic Chemicals, Bacteriacides, and Other Biocides is BAD Advice according to the US EPA, the American Industrial Hygiene Association, and the California Department of Health Services.

Bleach is ineffective and mold will begin to re-establish itself in less than 24 hours after applying bleach to any non-porous surface.

Mold spores are not visible to the human eye. They are microscopic. There can be 10,000 spores per cubic meter in the air in a room, and you would not see them. **Buying a house or any building without getting a mold inspection that includes collection of both indoor-outdoor air samples makes the purchase a HIGH RISK endeavor.**

Appendix A

If there is a mold problem, insurance companies will not pay for cleaning it up, you will. And if you make the mistake of reporting mold problems to your insurance company and filing a claim, your property will be listed in a national database that will affect insurability and the appraised value of your property forever.

All fungi can cause health problems for both humans and pets, according to a 2004 University of Connecticut Health Center Report.

Most Standard and Conventional Treatment Options Do Not Work and Do Not Provide Long-Term Protection.
Standard biocides and fungicides are so toxic they have been created to dissipate in a relatively short period of time, and they cannot be used on a long-term basis or as a preventative because they might kill you as well as the mold.

The Remediator's "Tear It All Out" Wisdom is COSTLY Advice.
Removing and replacing all water-damaged and mold-infested materials before treating them with the non-toxic fungicidal agent discovered by Dr. Close can be very costly. It may also spread mold spores throughout a building and may well eliminate any potential for identifying the source of the mold problem.

When a property has a mold problem, the property owner will probably bear the cost to clean it up. Many properties are condemned every year due to toxic mold infestations.

So, what can you use? What can eliminate mold in as little as 24 hours and keep it from coming back?
A blend of therapeutic grade essential oils used in accordance with the 10 Step Essential

Oil Protocol was found to be far more effective at eliminating mold than any other treatment currently available, with an overall spore removal efficiency of 96.65%, 100% spore removal efficiency for seventeen species of mold, and long-term residual effects. Case-study participants reported significant health improvements following this protocol.

This blend contains Lemon, Cinnamon, Cloves, Eucalyptus, and Rosemary. It is non-toxic and approved by the FDA for human consumption. There is scientific evidence that this same blend of therapeutic grade essential oils also kills bacteria and viruses (Chao et. al., _Journal of Essential Oil Research_, 1998).
No other mold-treatment option available can say that. No other fungicide, bleach, or mold-remediation technology offers such a cost-effective means to eliminate and prevent mold, while protecting our health.

This blend of essential oils, used according to the 10-Step Protocol, is the only mold-treatment option available that can be used as a preventative in the home or office building for long periods of time. And it smells great.

This blend of essential oils does not just kill mold spores, it actually removes mold spores, both living and dead, from the air. It provides a safe, eco-friendly, means of protection for your family, your property, and your pets.

Following are two graphs that show data Dr. Close collected at two of the case-study sites, and the 10-Step Protocol he developed for preventing and eliminating mold.

The person presenting this information can assist you in acquiring the essential oils used in the tests performed by Dr. Close, or visit www.MoldRx4u.com.

The Non-Toxic Solution to Toxic Mold

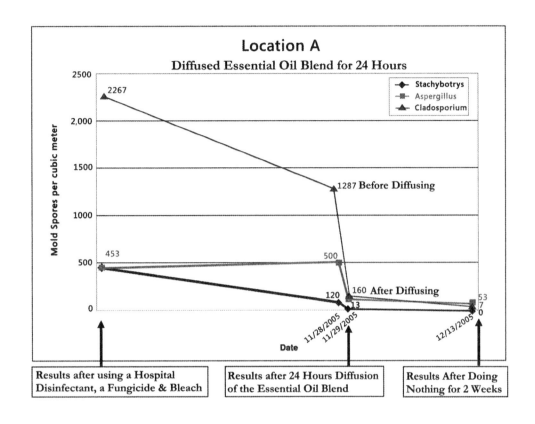

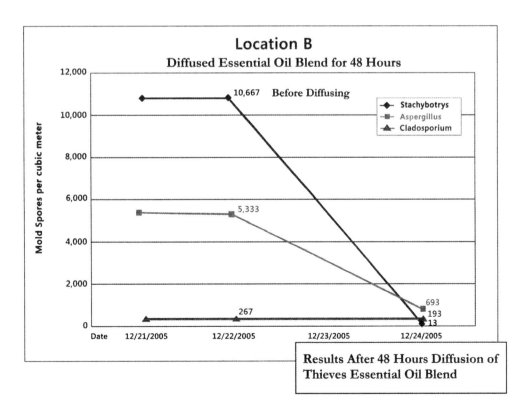

10 Steps to Prevent and Eliminate Mold
The Close Protocol (Patent Pending)

A. If you do <u>not</u> have mold -- PREVENT MOLD

At least once a week, diffuse the Thieves blend of essential oils for 15 minutes every 3 hours or continuously for 8 hours .

B. If you <u>have</u> mold -- ELIMINATE MOLD AND KEEP IT FROM COMING BACK

1. SAMPLE -- Before diffusing, have samples collected by a professional to determine the type of mold, and whether it is toxic or not. Toxic mold must be dealt with differently than non-toxic mold. DO NOT diffuse Thieves oil before samples are collected.

2. DIFFUSE -- After samples have been collected, eliminate mold spores by continuously diffusing the Thieves Oil blend for 24-72 hours in the space(s) where mold was found. One (1) Young Living cold-air diffuser works well in spaces up to 1000 sq. ft. in size. If possible, and for best results, leave the room closed and sealed during this intensive diffusing. This will allow maximum penetration and absorption of the essential oil blend. As a safety precaution, it is not recommended that pets or humans be in the space during this intensive diffusing.

3. REPAIR LEAKS -- Repair all leaks and eliminate all sources of excess moisture.

4. CLEAN THOROUGHLY -- If sampling identifies <u>NO toxic mold</u>, then after diffusing has been completed, clean visible mold and stains with Thieves Household Cleaner, undiluted. Use protective equipment and take precautions to avoid contact with and breathing mold spores while cleaning.

5. If sampling identifies <u>TOXIC MOLD</u>, and you have not already done so, begin intensive diffusion of Thieves oil. Contact a professional mold remediation service to have water-damaged and mold-infested materials removed and properly disposed.

6. REMOVE, REPAIR, and REPLACE all affected areas. If sampling revealed no toxic mold, remove water – or mold-damaged materials, seal them in plastic for disposal. Non-toxic materials may be placed in normal garbage disposal containers. If sampling identified toxic mold, be sure to diffuse continuously for a minimum of 24 hours following removal of mold-infested materials.

7. RE-SAMPLE - Have your professional resample to be sure all sources of mold have been identified and remediated.

8. REPEAT -- If necessary, repeat steps 1 through 7.

9. REFINISH all affected areas.

10. DIFFUSE REGULARLY to prevent and protect against the recurrence of mold. (See Item A, above, for details.)

APPENDIX B

A REPRESENTATION OF A LABORATORY REPORT
AND
CHAIN OF CUSTODY FORMS

XYZ Laboratories

Lab Address

Type of Sample and Type of Report; Lab method and Operating Procedure

Client Name: EJC Close and Address			Date of Report: 7/12/07	
Lab Sample ID No.:	XXXXXXX-01	XXXXXXX-02	XXXXXXX-03	
Client Sample ID:	ST -1	ST-2	ST-3	
Location:	Outside	Office 101	Hallway	
Project:	ABC Offices	ABC Offices	ABC Offices	
Collection Date:	7/9/2007	7/9/2007	7/9/2007	
Date Received by Lab:	7/12/2007	7/12/2007	7/12/2007	
Date Analyzed:	7/12/2007	7/12/2007	7/12/2007	
Parameter	Raw Counts Spores/m3	Raw Counts Spores/m3	Raw Counts Spores/m3	

Parameter	Raw Counts	Spores/m3	Raw Counts	Spores/m3	Raw Counts	Spores/m3
Volume (liters)	150		150		150	
Background Debris*	Light		Heavy		Moderate	
Limit of Detection (spores)	<1		<1		<1	
Alternaria			2	13	1	7
Asceospores	86	573	2	13		
Basidiospores	245	1,633	14	93	4	27
Bipolaris/Drechslera			1	7	2	13
Botrytis						
Chaetomium			5	33	3	20
Cladosporium	4	27	3116	20,773	80	533
Curvularia			4	27		
Epicoccum	1	7				
Fusarium						
Nigrospora			2	13		
Other Colorless	2	13				
Other Brown	2	13	2	13	1	7
Pennicillium/Aspergillus	36	240	33	220	20	133
Rusts			1	7		
Smuts, Myxomycetes, etc.	2	13				
Stachybotrys chartarum			3	20		
Stemphylium			1	7		
Torula			2	13		
Trichoderma			1	7		
Other:						
Total Spores/m3		2,519		21,252		740

Note: The data on this page are typical of lab reports for the Mississippi Valley Area, but do not represent actual analytical data from any real laboratory.

Appendix B

XYZ Laboratories   Lab Address

Type of Sample and Type of Report; Lab method and Operating Procedure

Client Name: EJC Close   Date of Report: 7/12/07
and Address

	XXXXXXX-01			XXXXXXX-02		
Lab Sample ID No.:	XXXXXXX-01			XXXXXXX-02		
Client Sample ID:	TL -1			TL-2		
Location:	AC Vent (101)			Hallway		
Project:	ABC Offices			ABC Offices		
Collection Date:	7/9/2007			7/9/2007		
Date Received by Lab:	7/12/2007			7/12/2007		
Date Analyzed:	7/12/2007			7/12/2007		
Parameter	Sample CTS	CFU	LOD	Sample CTS	CFU	LOD
Sample Amount	1			1		
Medium Used	MEA			MEA		
Dilution Factore Plated	1 : 10, 1 : 100			1 : 10, 1 : 100		
Acremonium						
Alternaria						
Asperguillus flavus	2	2,000	<1000			
Asperguillus fumagatus				1	1,000	<1000
Asperguillus glaucus						
Asperguillus nidulans						
Asperguillus niger	4	4,000	<10	14	14,000	<1000
Asperguillus ochraceus						
Asperguillus sydowii	1	1,000	<100			
Asperguillus versicolor				12	12,000	<1000
Basidiomycetes	14	14,000	<100			
Bipolaris/Drechslera						
Botrytis	2	2,000	<10			
Chaetomium	3	3,000	<10			
Cladosporium	4	4,000	<10			
Curvularia						
Epicoccum						
Mucor	2	20	<10			
Non-sporulating fungi	1	1,000	<1000	1	1,000	<10
Paecilomyces						
Penicillium	4	4,000				
Phoma/coelomycetes						
Rhizopus						
Stachybotrys chartarum	45	45,000	<100	12	12,000	<100
Ulocladium						
Yeasts	3	300	<10			
Total CFU/Swab		78,320			40,000	

Note: The data on this page are typical of lab reports for the Mississippi Valley Area, but do not represent actual analytical data from any real laboratory.

Appendix B

XYZ Laboratories Address Phone Nos.							COC Page __ of __								
Chain of Custody Form							Analyses								
EJC Enterprises P.O. Box 368 Jackson, MO 63755		Alternate billing information													
Report to: Ed Close		Email to:													
Project Description: Mold sampling, ABC Office Building															
Phone: Fax:		Client Project #													
Collected by (signature)		Turnaround Rush?, Next Day, Two Day, Date Needed:													
Sample ID	Sample Description	Type	Vol.	Date	Time		DIRECT EXAM	QUANTITATIVE FUNGAL	CULTURABLE AIR FUNGI (ANDERSON)	QUANTATIVE BACTERIA	CULTURABLE AIR BACTERIA	E. COLI / COLIFORM (presence/absence)	ENTEROCOCCUS (presence/absence)		
ST - 1	Outside	ST	150 L	120707	2:15 pm		X								
ST - 2	Office #101	"	"	"	2:35 pm		X								
ST - 3	Hallway	"	"	"	2:55 pm		X								
TL - 1	AC Vent	Tape	NA	"	3:15 pm			X							
TL – 2	Hallway	"	NA	"	3:45 pm			X							

Type = **Tape –** Tapelift, **Bulk -** Bulk, **Swab -** Swab, **ST –** Spore Trap, etc., **AF –** Anderson Fungal. **AB – Anderson Bacterial**

Comments:

Relinquished by: (Signature)	Date	Time	Received by: (Signature)	Shipped Via:	Condition:	
Relinquished by: (Signature)	Date	Time	Received by: (Signature)	Temp.		
Relinquished by: (Signature)	Date	Time	Received by: (Signature)	pH	NCF	

APPENDIX C

HOSPITALS USING ESSENTIAL OILS

There is ample evidence that plant materials have healing properties and have been used by human beings for healing for centuries. Now, numerous research studies, many cited in this book, prove that various essential oils from plants have significant antimicrobial properties. This being the case, why aren't they being widely used by doctors and hospitals? The truth is that they are being used in other parts of the world, especially in France, where the use of essential oils is an accepted and established specialty in medical practice.

We could speculate about the reasons essential oils are rarely used in U.S. hospitals. Is it because we are so enamored with modern technology that we overlook the fact that synthetic medicines always produce negative side effects? Or does the close relationship among doctors, hospitals, and the multibillion-dollar pharmaceutical industry have something to do with it? Is it a choice, a group of choices made in the past, a widely held belief system, a conspiracy? Many people have opinions; however, it is likely that a combination of many factors has brought us to the point where we are today. In our rush to seek out new answers, and in our belief that current science and technology are always correct and that new is always best, we can overlook the simple solutions to many problems that are simply right under our noses.

Fortunately, there are some signs that attitudes are changing. More and more doctors are interested in integrated, complimentary, and alternative medicine. Some hospitals, hospices, and nursing homes are using essential oils to deodorize their facilities, calm their patients, and reduce cross infections. The following are just a few examples.

In **Worchester Hospital in Hereford, England** they did a six-month study in which they discovered that vaporizing lavender caused their patients to have more natural sleep patterns and made them less aggressive. Many patients were able to be weaned of tranquilizers altogether.

In **Minneapolis**, essential oils are used to reduce the wandering of elderly patients.

At **Churchill Hospital in Oxford, England,** many of the Alzheimers patents treated with essential oils have become more alert. Patients with dementia have become calmer.

St. Croix Valley Memorial Hospital in Wisconsin uses essential oils throughout the lobby, at the nurses' stations, and the emergency waiting room. Anxiety relieving essential oils are used. The hospital also has two floater mobile units to be used whenever a special need arises.

174

At **Memorial Sloan-Kettering Cancer Center in New York**, they tested the anxiety level of patients going through MRI. Forty-two patients breathed normal air, and thirty-eight breathed air with essential oils; 63% of those exposed to the aromatic oils experienced reduced anxiety levels.

At **St. John's and St. Elizabeth's Hospital in London,** most of the midwives have become trained aromatherapists. They use essential oils from the beginning of pregnancy through the aftercare.

At **Royal Sussex County Hospital** in Brighton, England, thirty-six patients in the intensive-care and coronary-care units were tested to determine whether there was any benefit to using essential oils combined with massage. The control group received nothing, a second group received massage without essential oils, and a third group received aromatherapy massage. The patients' progress was followed over a five week period.

Systolic blood pressure (the first number of a blood-pressure reading) dropped 50% for those who received aromatherapy with massage, 40% with massage alone, and 16% for the control group. Respiratory rates decreased by 75% for the aromatherapy group, 41% for massage alone, and 16% in the control group. Heart rate decreased by 91% for the aromatherapy group, 58% for those receiving massage alone, and 41% for the control group.

APPENDIX D

MSDS SHEETS

FOR THE

THIEVES® HOUSEHOLD CLEANER

AND THE

INDIVIDUAL OILS CONTAINED IN THE THIEVES® ESSENTIAL OIL BLEND

Thieves®
HOUSEHOLD CLEANER
CLEANING AGENT
Item # 3743
Formula Number 1B
Vendor:
Contact:

Phone:
Fax:

INGREDIENTS	
Energy 2	98% of formula
Liposomes 15%	
Essential Oil Blend - Thieves	
Clove EO (*Syzygium aromaticum*)	
Lemon EO (*Citrus limon*)	
Cinnamon Bark EO (*Cinnamomum verum*)	
Eucalyptus Radiata EO (*Eucalyptus Radiata*)	
Rosemary EO (*Rosmarinus officinalis* CT 1.8 *Cineol*)	

Material Safety Data Sheet   Date Issued: 1 – 31 – 05

SECTION I   **IDENTIFICATION OF PRODUCT**

Product Name: ENERGY II SUPER CONCENTRATED CLEANER
Description: A SUSPENSION AGENT, NORMALLY DILUTED WITH WATER

SECTION II   **HAZARDOUS INGREDIENTS**

Contains no hazardous ingredients under current OSHA definitions

This material mixture formula is a trade secret and complies with 29CFR XVIII-1910.1200 SECTION (I) TRADE SECRETS.

This formula contains no ingredients that are on the NTP list or registered with IARC for carcinogens and the material mixture as a whole has been found to be:

1)NON TOXIC 2) NOT CORROSIVE 3) NOT AN IRRITANT 4) NOT A SENSITIZER IN ORAL, DERMAL, AND OCULAR TESTS ON ALBINO RATS AND RABBITS PER FEDERAL HAZARDOUS SUBSTANCES ACT (15CFR 1500)

Appendix D

SECTION III   **PHYSICAL DATA**

Boiling Point at 1 ATM	
DEG C:	93.4
Vapor Pressure 20C	
(MMHG):	17.5
Vapor Density (Air 1)	.62
Solubility in water:	Complete
Appearance and Odor:	Light Yellow-Mild Odor
Specific Gravity (H_2O=1)	1.010 at 20 DEG F.
Evaporation Rate:	.7
PH Range (Concentrate)	10.7

SECTION IV   **FIRE AND EXPLOSION DATA**

Flash Point: Non-flammable; decomposed at 110C without flame using Pensky-Martens closed tester 211.7 – 1979

Flammable Limits: LEL N/A UEL – N/A

Extinguished Media: OK with CO_2, Sand, Dry Chemicals, Foam, Water, Spray.

Special Fire Fighting
Procedures: Poses no unusual fire or explosive hazards

SECTION V   **HEALTH AND HAZARD DATA**

EYE CONTACT: In concentrated form irritation may occur in some cases if splashed in eye. Flush with water as one would do with regular shampoo.

INHALATION: No health effects are known to occur from inhalation of this material under normal use in a normally ventilated area.

INGESTION: Ingestion of concentrate is not recommended but no special precautions are needed. If small amounts are consumed, drink water to get rid of the taste.

SECTION VI   **SPILL, LEAK AND DISPOSAL PROCEDURES**

Spill Response: Areas may become slippery, flush with water to dilute. Spilled product may be disposed of down any drain to clean the drain.

Product Disposal: May be flushed to any sewer system for normal disposal.

Container Disposal: Rinse used container with water and discard in regular trash.

Other Precautions: If exposed to low temperatures (below 32F) product may stratify. If this happens, mix before using.

SECTION VII   **CONTROL MEASURES**

No precautionary measures needed. Any normal ventilation is adequate. Eye protection or gloves are not necessary. As with any cleaner, keep out of reach of children.

THIS MSDS IS IN COMPLIANCE WITH THE OSHA HAZARD COMMUNICATION REGULATIONS. ALL THE INFORMATION CONTAINED ON THIS MSDS IS THE LATEST UNDER NORMAL CONDITIONS. FOLLOW INSTRUCTIONS ON THE LABEL AND INSTRUCTION SHEETS. ANY USE OR METHOD OF APPLICATION NOT IN CONFORMANCE WITH THIS MSDS, PRODUCT LABEL OR INSTRUCTION SHEETS IS THE TOTAL RESPONSIBILITY OF THE USER.

Material Safety Data Sheet   11.03.2003

1.   IDENTIFICATION OF SUBSTANCE

Product Name:	CINNAMON BARK ESSENTIAL OIL
Botanical Name	*Cinnamomum zeylanicum*
FEMA No.	2291
CAS Number:	8015-91-6

2.   INFORMATION ON INGREDIENTS

Pure essential oil

3.   HAZARDOUS IDENTIFICATION

Most important hazards:
Harmful in contact with skin
May cause sensitization by skin contact

4.   FIRST AID MEASURES

Inhalation: In case of excessive inhalation remove person to fresh air; keep at rest & comfortable. Obtain medical advice immediately.

Skin Contact: Remove contaminated clothing (N.B. wash before reuse). Wash off of skin immediately with plenty of water, using soap if available. If any sign of tissue damage or persistant irritation is apparent obtain medical advice immediately.

Eye Contact: Rinse eyes immediately with plenty of water for at least ten minutes. If any sign of tissue damage or persistant irritation is apparent obtain medical advice immediately.

Ingestion: Rinse mouth wit water. Obtain medical advice immediately.

5.   FIRE-FIGHTING MEASURES

Extinguishing Media: Carbon dioxide; Foam; Dry chemical
Do not use a direct water jet.

Combustion Products: CO/CO2, Smoke

Special Protective Equipment:
Avoid inhalation of smoke & fumes. Wear suitable respiratory equipment.

6.   ACCIDENTAL RELEASE MEASURES

Personal precautions: Gloves and eye protection should be worn when handling spillages. Avoid flame or other potential sources of ignition. Avoid skin/eye contact and inhalation of vapour.

Appendix D

Good personal washing routines should be followed after accidental releases. Ensure adequate ventilation in working areas following accidental release.

Environmental Precautions: Do not allow discharge into drains, soil, or any aquatic environment.

Spillage: Absorb spillages on porous inert material such as earth, sand or vermiculite and dispose of in accordance with local regulations.

Large spillages should be contained by the use of sand or another inert material, transferred to a suitable container and recovered or disposed of in accordance with local regulations.

---

7.   HANDLING AND STORAGE

General Handling Precautions:
Handle in accordance with good occupational hygiene and safety practices in a well ventilated area. Avoid direct contact with skin and eyes. Depending on working conditions, this may include wearing of eye protection and protective clothing such as PVC gloves and suitable overalls. Avoid breathing vapours especially if the material is hot.

Storage Conditions:
Store in full, dry, airtight containers away from sources of heat and light. Do not re-use the empty container.

---

8. EXPOSURE CONTROLS

Respiratory protection: Where ventilation may be inadequate, wear self-contained breathing apparatus.
Hand protection: Wear impervious gloves.
Skin protection: Wear protective clothing.
Eye protection: Wear Eye protection e.g. safety glasses/goggles

---

9. PHYSICAL CHEMICAL PROPERTIES

Appearance: Golden brown liquid
Odour: Warm spicy cinnamon
RI @ 20°C: 1.5520 – 1.5660
OR @ 20°C: 1° to – 1°
SG @ 20°C: 1 to 1.040
Solubility in 70% ethanol: 3 to 1 (slight opalescence)
Aldehyde content (calculated: Not less than 40%
As cinnarnic aldehyde) Typical properties: Flash Point 71° C

---

10.   STABILITY AND REACTIVITY
Stability: Stable under normal conditions

Incompatible Materials: Avoid strong oxidizing agents

180

11. TOXICOLOGICAL INFORMATION

Acute oral toxicity: LD50:   (Rat)   3.4 mg/Kg
Acute dermal toxicity: Moderately irritant
Sensitization: Sensitizing
Phototoxicity   Non phototoxic

12. ECOLOGICAL INFORMATION

No data available
Prevent contamination of soil, ground and surface water.

13. DISPOSAL CONSIDERATIONS

No special methods are necessary, but disposal should be in accordance with local regulations.

14. TRANSPORT INFORMATION

Not restricted for transport purposes

15. REGULATORY INFORMATION

Classified as HARMFUL
Symbol: Xn

R phrases:
R21: Harmful in contact with skin
R43: May cause sensitization by skin contact

S-phrases:
S24/25: Avoid contact with skin and eyes

16. OTHER INFORMATION

None

Appendix D

CLOVE BUD ESSENTIAL OIL

MSDS   Stamped Received Oct 15 2002

1. IDENTIFICATION :--

PRODUCT NAME: Clove Bud essential oil
BOTANICAL NAME: *Eugenia caryophyllata*
FEMA #: 2323
FDA #: 184.1257
CAS #: 80000-34-8
MAIN COMPONENT: Eugenol, eugenyl acetate, beta-caryophyllene
REGULATORY STATUS: GRAS (I), (II)

2. PHYSICAL DATA:--

APPEARANCE: Pale yellow liquid
ODOUR: Spicy clove-like aroma
OPTICAL ROTATION: -0.25 deg.
FLASH POINT: >212 deg. F
SPECIFIC GRAVITY: 1.052
BOILING POINT: >150 deg. C
REFRACTIVE INDEX: 1.5310
MELTING POINT: <0 deg C
SOLUBILITY IN ETHANOL: Soluble

3. FIRE, EXPLOSION AND REACTIVITY:--

EXTINGUISHING MEDIA: Carbon dioxide; Foam; Dry chemical
SPECIAL FIRE-FIGHTING PROCEDURES: Self-contained breathing apparatus and protective clothing should be worn if chemical used
UNUSUAL FIRE & EXPLOSION HAZARD: Nra
HAZARDOUS POLYMERIZATION: Not known to occur
MATERIAL TO AVOID: Avoid strong oxidizing agents
CONDITION TO AVOID: Avoid strong oxidizing agents
HAZARDOUS DECOMPOSITION PRODUCTS: None
CHEMICALLY STABLE: Yes

4. SPILL, LEAK AND DISPOSAL PROCEDURES:--

PRECAUTION IF MATERIAL IS SPILLED OR RELEASED: Small spills can be wiped up with cloth or absorbents can be used; standard absorbents can be (sand, sawdust, vermiculite, etc.) Avoid open flames or other sources of ignition.
WASTE DISPOSAL METHOD: Cover with an inert non-combustible absorbent material and remove to approved disposal container. Dispose of in accordance with local, state and federal regulations.

Appendix D

5. PROTECTION INFORMATION:--

VENTILATION: Provide adequate ventilation. Use local exhaust fan if necessary.
RESPIRATORY PROTECTION: Usually not required in well-ventilated areas. In a confined, poorly-ventilated area, use of "NIOSH" approved respiratory protection may be required.
EYE PROTECTION: Safety glasses recommended. The use of splash goggles is recommended if splashing occurs.
SKIN PROTECTION: Chemical resistant gloves recommended.

6. HEALTH HAZARD DATA:---

OCCUPATIONAL EXPOSURE LIMIT: N/A
THRESHOLD LIMIT VALUE (TLV): N/A
OSHA PERMISSIBLE EXPOSURE LIMIT (PEL): N/A
HEALTH HAZARD DETERMINATION: Liquid is irritating to skin and eyes
EFFECTS OF OVEREXPOSURE: N/A
EYE CONTACT: Irrigate with water at least 15 minutes. obtain medical advice immediately..
SKIN CONTACT: Remove contaminated clothes, wash affected area with water, if irritation persists, obtain medical advice.
INHALATION: Leave area of exposure to well-ventilated area or fresh air. Contact a physician if necessary.
INGESTION: If accidental ingestion occurs, rinse mouth with water. Give up to one half pint of milk or water. Obtain Medical advice immediately.

7. TRANSPORTATION INFORMATION:---
Non- hazardous material transport in suitable containers.

8. STORAGE AND HANDLING:--
It is recommended that this material be stored in tight containers in a cool, dry place away from light and possible sources of ignition.

PREPARED BY:

MATERIAL SAFETY DATA SHEET

1. IDENTIFICATION OF SUBSTANCE/PREPARATION & COMPANY.
PRODUCT NAME: NARROW-LEAVED PEPPERMINT EUCALYPTUS ESSENTIAL OIL – RADIATA
SUPPLIER:

2. COMPOSITION/INFORMATION ON INGREDIENTS.
DEFINITION/BOTANICAL ORIGIN: Obtained by steam distillation from the leaves & twigs of *Eucalyptus radiata*

H-No: 3301.29	CAS Number: 8000-48-4	Status: natural
CoE No:	FEMA:	Additives: none
FDA:	IFRA:	Application: Cosmetics, toiletries, flavours
RIFM:	EINECS:	FCCIV: official
INCI name: Eucalyptus radiate		

3. HAZARDS IDENTIFICATION:
Xn. Harmful

4. FIRST-AID MEASURES:
INHALATION: Remove from exposure site to fresh air. Keep at rest. Obtain medical attention.
EYE CONTACT: Rinse immediately with plenty of water for at least 15 mins. Contact a doctor if symptoms persist.
SKIN CONTACT: Remove contaminated clothes. Wash thoroughly with soap and water, flush with plenty of water. If irritation persists, seek medical advice.
INGESTION: Rinse mouth out with water. Seek medical advice **immediately**.
OTHER: When assessing action take Risk & Safety Phrases into account (Section 15)

5. FIRE-FIGHTING MEASURES.
EXTINGUISHING MEDIA RECOMMENDED: Use CO_2, Dry Powder or Foam type of Extinguishers, spraying extinguishing media to base of flames. Do not use direct water jet on burning material.

SPECIAL MEASURES: Avoid vapour inhalation. Keep away from sources of ignition. Do not smoke. Wear positive pressure self-contained breathing apparatus and protective clothing.

EXTINGUISHING PROCEDURES: Closed containers may build up pressure when exposed to heat and should be cooled with water spray.

6. ACCIDENTAL RELEASE MEASURES.
PERSONAL PRECAUTIONS: Avoid inhalation & direct contact with skin & eyes. Use individual protective equipment (safety glasses, waterproof-boots, suitable protective clothing) in case of major spillages.

ENVIRONMENTAL PRECAUTIONS: Keep away from drains, soils, surface and groundwaters.

CLEANING UP METHODS FOR SPILLAGES: Remove all potential ignition sources. Contain spilled material. Cover with an inert or non-combustible inorganic absorbent material, or sweep up and remove to an approved disposal container.

7. HANDLING AND STORAGE:
PRECAUTIONS IN HANDLING: Apply good manufacturing practice and industrial hygiene practices, ensuring proper ventilation. Observe good personal hygiene, and do not eat, drink or smoke whilst handling.

Appendix D

STORAGE CONDITIONS: Store in tightly closed original container, in a cool, dry & ventilated area away from heat sources & protected from light. Keep air contact to a minimum.

FIRE PROTECTION: Keep away from ignition sources & naked flames. Take precautions to avoid static discharges in working area.

8. EXPLOSION CONTROLS/ PERSONAL PROTECTION

RESPIRATORY PROTECTION:Avoid breathing product vapour. Apply local ventilation where possible.

VENTILATION: Ensure good ventilation of working area.

HAND PROTECTION: Avoid all skin contact. Use chemically resistant gloves if required.

EYE PROTECTION: Use safety glasses.

WORK/HYGIENE PRACTICES: Wash hands with soap & water after handling.

9. PHYSICAL AND CHEMICAL PROPERTIES:

COLOUR: Clear colourless to pale yellow
APPEARANCE: Clear mobile liquid
ODOUR: Characteristic of peppermint
FLAVOUR: Characteristic of peppermint
FLASH POINT °C: 48
RELATIVE DENSITY: d_{20}^{20}: 0.8850 to 0.9260
OPTICAL ROTATION @ 20° C: Not available
REFRACTIVE INDEX @ 20° C: Not available
SOLUBILITY IN VEGETABLE OILS:
SOLUBILITY IN WATER: Negligible
SOLUBILITY IN 85% ETHANOL: Soluble
ADDITIONAL DATA:
TYPICAL ANALYSIS BY GC/MS: 1,8-Cineol >70% TYPICAL, Terpene hydrocarbons <15% TYPICAL, Other Components< 15%

10. STABILITY AND REACTIVITY

REACTIVITY: It presents no significant reactivity hazards by itself, or in contact with water. Avoid contact with strong acids, alkali or oxidizing agents.

DECOMPOSITION: Liable to cause smoke & acrid fumes during combustion: carbon monoxide, carbon dioxide & other non-identified organic compounds may be formed.

11. TOXOLOGICAL INFORMATION
No data available

Material Safety Data Sheet

I. PRODUCT IDENTIFICATION

Tradename: Cold Pressed Lemon Oil
Product Code: 340000
CAS Number: 84929-31-7

II. HAZARDOUS INGREDIENTS

Materials: N/A

III. EMERGENCY CONTACT

*ONLY IN THE EVENT OF CHEMICAL EMERGENCIES INVOLVING A SPILL, LEAK, FIRE, EXPOSURE, OR ACCIDENT INVOLVING CHEMICALS**

CHEMTREC
United States Shipments **(800) 424-9300**   Outside United States **Call CHEMTREC Collect**

IV. PHYSICAL DATA

Specific Gravity: 0.852 @ 20/20° C
Appearance: Yellow to pale yellow
Odor: Odor and flavor similar to sweet lemon
Solubility in water: Insoluble
Refractive Index: 1.472 – 1.476
Optical Rotation: 60 – 70 degrees
Aldehyde: 2.0 – 3.5%
Density: 0.847 – 0.853

V. FIRE AND EXPLOSION HAZARD DATA

Flashpoint: EOA Closed cup 116° F
Extinguishing Media: Class B; Smothering or exclusion of air.
Unusual Fire & Explosion Hazards: Burns with intense heat, container may explode in heat of fire.

VI. HEALTH AND HAZARD DATA

Effects of Overexposure: Skin irritant
Emergency First Aid & Procedures: Flush eyes with large quantities of water, rinse other areas (when appropriate) with soap and water and apply petroleum jelly if necessary; consult physician if required.

Stamped Received Feb. 2 2006

Appendix D

MATERIAL SAFETY DATA SHEET

1. IDENTIFICATION OF THE PRODUCT AND THE COMPANY

Commercial name: **ROSEMARY ESSENTIAL OIL**
Supplier
Emergency Telephone   Consult the toxicological institute of your country

2. COMPOSITION / INFORMATION ON COMPONENTS

Common name   ROSEMARY ESSENTIAL OIL
                     Rosmarinus officnalis L.
Nature of product: 100% pure and natural
CAS TSCA: 8000-25-7
CAS EINECS   84604-14-8

Components contributing to
the danger

3. HAZARDS IDENTIFICATION

Main effects on health   Flammable and harmful
and environment

4. FIRST-AID MEASURES

Skin contact: Remove contaminated clothing. Rinse affected area with plenty of water. Contact a physician if
                  symptoms persist.
Eye contact: Flush with water for at least 15 minutes. Contact a physician if symptoms persist.
Ingestion: Rinse mouth out with water and seek medical advice immediately.
Inhalation: If discomfort is felt after inhalation of vapors, remove affected person to fresh air. In case of
                  breathing difficulty seek medical advice.

5. FIRE-FIGHTING MEASURES

Extinguishing Media
recommended: Foam, carbon dioxide or dry chemical.
Non-recommended: Direct water jet
Special Precautions: Toxic vapors may be released during fire. Respiratory protection should be provided.
                     Containers close to the fire should be cooled with water spray.

6. ACCIDENTAL RELEASE MEASURES

Individual protection: Avoid eye and skin contact. Prevent contact with hot surfaces.
Environment protection: Keep away from drains, soil, surface- and groundwater.
Methods for cleaning up: Contain spilled material, cover with an inert non-combustible, inorganic, absorbent
                            material.

Appendix D

7. HANDLING AND STORAGE

Handling: Apply according to good manufacturing and industrial hygiene practices with proper ventilation. Use individual protective equipment. Do not drink, eat or smoke while handling.

Storage: Store in cool, dry and ventilated area, away from heat sources, sparks and open flames, protected from light in tightly closed original containers.

8. EXPOSURE CONTROLS AND PERSONAL PROTECTION

Respiratory protection: Avoid breathing fumes. Provide adequate ventilation.
Eye protection: Use safety glasses.

Hand protection: Avoid skin contact. Use protective gloves.

Skin Protection: Use protective clothing.

9. PHYSICAL AND CHEMICAL PROPERTIES

Appearance:	Mobil liquid limpid
Colour	Colourless to very pale yellow
Odour:	Strong, rustic, cineol note
Flash Point:	43 deg. C
Specific Gravity at 20°C	0.907 to 0.920
Refractive index at 20°C	1.464 to 1.470
Optical Rotation at 20°C	-2° to +5°
Solubility	

10. STABILITY AND REACTIVITY

Stability: No significant modifications in normal storage conditions.
Reactivity: No significant reactivity hazards in normal use conditions.
Conditions to Avoid: High temperature
Products to Avoid: Avoid contacts with strong acids, alkali or oxidizing agents

11. TOXICOLOGICAL INFORMATION

Effects by ingestion: LD 50 for the rat >5 ml/kg

Effects by skin contact: LD 50 for the rabbit > 10 ml/kg

FIFRA Restriction: None

12. ECOLOGICAL INFORMATION

Prevent contamination of soil, ground and surfacewater. Biodegradable product.

13. DISPOSAL CONCIDERATIONS

In accordance with local environmental laws. Avoid disposing into drainage systems and into the environment.

14. TRANSPORTATION REGULATIONS

Customs code   33 01 29 61

TRANSPORT   Class 3 (Flammable Liquid) UN 1169

15. REGULATORY REGULATIONS

Labelling   Xn (harmful), R10 (flammable), R65
(can provoke lung problem in case of ingestion), S62
(do not induce vomiting in case of ingestion,
see immediately a doctor and show him the label or the packaging).

16. OTHER INFORMATION

The information in this safety data sheet is based on the properties known to ALBERT VIEILLE S.A. at the time the data sheet was issued. The safety data sheet is intended to provide information for a health and safety assessment of the material and the circumstances under which it is packaged, stored or applied in the workplace. For such safety assessment ALBERT VIEILLE S.A. holds no responsibility. This document is not intended for quality assurance purposes.

          Issue: 22/04/1994   Revision :   11/04/2005

GLOSSARY

Adulterated essential oil: An oil that has had synthetic compounds added to the natural essential oil.

AFNOR: Association française de normalisation (the French association for the normalization of environmental regulations).

Aromatherapy: The use of volatile liquid plant materials known as essential oils for the purpose of improving a person's mental and physical health.

Aromatherapist: A trained and qualified professional who practices and/or teaches aromatherapy.

Antibacterial: Able to destroy or inhibit the growth of bacteria.

Antifungal: Able to destroy or inhibit the growth of fungi.

Antimicrobial: Able to destroy or inhibit the growth of micro-organisms.

Antiseptic: A substance that is able to destroy or prevent the development of micro-organisms.

Atypical Fungi: Fungi not normally found in a given environment.

Bactericide: A substance that kills bacteria.

Black Mold: A mold that is black or dark in appearance.

Black Toxic Mold: A black mold that produces a toxin or toxins.

Bleach: A chemical compound designed to remove or lighten color; usually sodium hypochlorite, calcium hypochlorite, peroxide, or a borate compound.

Bulk Sampling: Collection of a piece of substrate (e.g., sheetrock upon which mold is growing), carefully sealing it in a plastic bag, and submitting it to a laboratory for analysis.

Chain of Custody: The documentation or paper trail of the handling of samples or other physical evidence.

Chemotype (CT): Oils of the same species that have significantly different compositions in response to environmental factors.

Chlorine Bleach: An oxidizing agent with Sodium hypochlorite (NaClO) as its active ingredient, produced by infusing sodium hydroxide with chlorine gas.

Diffuse: To saturate a space with an aerosol or micro-mist of an essential oil or essential oil blend.

Diffuser: An instrument or piece of equipment used to diffuse essential oils.

Diffuser, cold-air: A diffuser that diffuses essential oils by means of a stream of air without heat or water.

Essential Oil (EO): Oil extracted from plant material by distillation.

EOB2: Essential Oil Blend #2; a proprietary blend of the essential oils of Clove (*Syzygium aromaticum*), Cinnamon Bark (*Cinnamomum verum*), Lemon (*Citrus limon*), Rosemary (*Rosmarinus officinalis CT cineol*), and Eucalyptus (*Eucalyptus radiata*) used in the case studies reported in this book.

FIFRA: The Federal Insecticide, Fungicide and Rodenticide Act.

Food-grade: Essential oil used as food additives for flavor, often augmented or adulterated with synthetic compounds.

Fungi: Members of a biological kingdom of organisms with cell walls constructed of chitin $(C_8H_{13}O_5N)n$ a long-chain polymeric polysaccharide of beta-glucose that forms a hard, semitransparent material that is also found in insects and crustaceans such as crabs and lobsters. In the majority of fungi species, these cells form multicellular filamentous growths called *hyphae* (branches) which form a *mycelium* (stalk). Sexual and asexual reproduction is via spores, produced on specialized structures called fruiting bodies. Molds, mildew, yeasts, and mushrooms are fungi. Some fungal species also grow as single cells.

Fungi, atypical: Fungi not normally found in a given environment.

Fungicide: A substance that kills fungi.

Fungal Infestation: The unwanted growth and spread of fungal colonies.

Fungus: The singular of *fungi*.

GRAS: Generally Regarded as Safe. A designation assigned by the US FDA, which applied to most essential oils in common usage.

Genus: A category in biological classification used to designate a group of closely-related species. For example, *Aspergillus* is a genus of mold comprised of nearly 200 species.

HVAC System: Heating, ventilation and air conditioning system.

HEPA Filter: High efficiency particulate air filter. Filters with this designation should remove at least 99.97% of airborne particles 0.3 micrometers or larger in diameter.

Hyphae: Multi-cellular filamentous growth forming the *mycelium* of a fungus.

IAQA: Indoor Air Quality Association.

ISHA: Institute of Spiritual Healing and Aromatherapy.

ISO: International Standards Organization.

LD50, LD$_{50}$: The dosage of a substance that results in killing 50% of lab animals used in a scientific study to determine the lethal dosage of a given substance.

Mildew: An almost microscopic fungus, usually white or gray, powdery in appearance that grows primarily on plants.

Moisture Barrier: Any material or structure designed to prevent water or water vapor from penetrating the living and/or storage spaces of a building.

Mycelium: The vegetative part of fungi, consisting of thread-like branches called *hyphae*.

Mycology: The discipline of biology devoted to the study of fungi.

Mycotoxin: Any toxin produced by a fungus. Molds produce mycotoxins from other chemicals like polypeptides and amino acids, which they use in metabolism.

Bibliography

MSDS: Material Safety Data Sheet required by OSHA for the occupational handling of toxic and hazardous substances in the workplace.

MVOC: Microbial volatile organic compounds responsible for musty or moldy odors that sometimes accompany mold infestations. MVOCs are derived from alcohols, ketones, and hydrocarbons. They are gaseous at room temperature and mix easily with air to impact our olfactory nerves

NAHA: The National Association for Holistic Aromatherapy. An educational non-profit organization dedicated to enhancing public awareness of the benefits of aromatherapy.

OSHA: US Department of Labor Occupational Safety and Health Administration, the government agency that establishes and enforces protective standards for people in the workplace.

PE (Professional Engineer): Any person with a degree in engineering or science, who, after working under the direct supervision of a registered professional engineer for a specific period of time, passes an eight-hour test administered by the state.

Perfume-grade: Essential oil used as soap or cosmetics additives for aroma, often augmented or adulterated with synthetic compounds.

Personal Protective Equipment (PPE): Respiratory equipment, clothing, and barrier materials used to protect the wearer from exposure to biological, chemical, and physical hazards. Different levels of PPE may be used depending on the hazard present.

Protocol: A specific routine or series of actions designed to achieve a specific purpose.

The **ten-step protocol** described in this book is a routine using essential oils for the purpose of eliminating the hazards of mold infestation.

RA: Registered Aromatherapist. Any person who has completed a minimum of one year of level 2 training in a college or school approved by NAHA, and successfully passes the Aromatherapy Registration Council's four-hour comprehensive test.

Raindrop Technique: A structures method developed by D. Gary Young for applying therapeutic-grade essential oils in a massage setting.

Remediation (mold): The elimination of mold growth, mold toxins and mold spores from indoor environments.

Spore Removal Efficiency (SRE): The percentage of available spores removed by a protocol designed to eliminate mold from an indoor environment.

Spore Trap Sampling: A method of determining the concentration (spores per unit volume of air) in the air by pulling a large volume of air through a small device containing a sticky microscope slide upon which the spores will impact and stick. In the laboratory, under a microscope, the spores are counted, segregated by species, and the number of each species per cubic meter of air is determined.

Standardized essential oil: Essential oils that have had synthetic compounds, or lesser grade essential oils, or a different species of plant added in order to alter the chemical composition of the oil and meet a certain standard. This is a common practice in the food and fragrance industry. **Note:** Standardized oils are not considered therapeutic quality, and are not suitable for therapeutic applications.

Still: An apparatus designed to distill the essential oils from plant material.

Synthetic Oil: An oil manufactured in the laboratory.

Species: A basic unit of biological classification designating organisms of the same genus. For example, *Aspergillus versicolor,* is a mold named *versicolor,* a member of the genus of molds known to mycologists as *Aspergillus.*

Tape Lift Sampling: A method of sampling utilizing clear tape to lift mold from an infested surface.

Therapeutic-Grade Essential Oil: An essential oil produced for therapeutic (or healing) purposes. The lipid (oil) soluble volatile, aromatic compounds obtained by steam distillation and cold expression or cold pressing of certified organic plant materials in a way that preserves the essential oil in a form that is as close to nature as the extraction process will permit, and that every reasonable effort has been made to prevent contamination by synthetics, chemicals, preservatives, and cleansers, in order to preserve the living energy and therapeutic benefits of the oil. **Note:** Perfume-grade and food-grade oils may contain substances that render them unsuitable for therapeutic applications, and unsuitable for mold remediation, prevention and elimination.

Toxic Mold: A mold that produces a toxin or toxins.

Toxicity: A measure of the degree to which a substance is poisonous and harmful to living organisms, especially human beings.

Toxicology: The study of poisons and their effects on living organisms.

Viable Sample: A sample containing living materials; e.g., mold spores that can be cultured to produce living specimens of the originating organism.

Wall Probe Sampling: A method of pulling air from inside wall cavities into a spore trap.

Bibliography

BIBLIOGRAPHY

List of Reference Books
Alphabetical by Title

1. **Advanced Aromatherapy, The Science of Essential Oil Therapy**, Kurt Schnaubelt, PhD, Healing Arts Press, 1998.

2. **Aromatherapy, A Complete Guide to the Healing Art**, Kathi Keville and Mindy Green, The Crossing Press, a division of Ten Speed Press, 1995.

3. **Aromatherapy, A Lifetime Guide to Healing with Essential Oils**, Valerie Cooksley, Prentice Hall, 1996.

4. **Aromatherapy For Health Professionals**, by Shirley Price and Len Price, Churchill Livingstone, an Imprint of Harcourt Publishers Limited, 1995-2001.

5. **Aromatherapy Scent and Psyche**, Peter & Kate Damian, Healing Arts Press, 1995.

6. **Aromatherapy: The Essential Beginning**, D. Gary Young, ND, Essential Press Publishing, 1996.

7. **Aromatherapy to Health and Tend the Body**, Robert Tisserand, Lotus Press, 1988.

8. **A Statistical Validation of Raindrop Technique**, David Stewart, PhD, RA, CARE Publications, 2003.

9. **Black Mold, Your Health and Your Home**, Richard F. Progovitz, The Forager Press, 2003.

10. **Chemistry of Essential Oils Made Simple, God's Love Manifest in Molecules**, David Stewart, PhD, DNM, CARE Publications, 2005.

11. **Clinical Aromatherapy, Essential Oils in Practice**, Second Edition, Jane Buckle, RN, PhD, Churchill Livingstone, an Imprint of Elsevier Limited, 2003.

12. **Essential Chemistry for Safe Aromatherapy**, Sue Clarke, Churchill Livingston, 2002.

13. **Essential Oils Desk Reference**, Third Edition, Complied by Essential Science Publishing © 2006, Third Printing March 2006.

14. **Essential Oils for Physical Health and Well-Being**, Linda L. Smith, RN, MS, CCA, HTSM Press, 2006.

15. **Essential Oils Integrative Medical Guide**, D. Gary Young, ND, Essential Science Publishing, 2003.

16. **Essential Oil Safety**, Robert Tisserand and Tony Balacs, Churchill Livingstone, an Imprint of Harcourt Publishers Limited, 1995-1999.

17. **Gattefosse's Aromatherapy**, Rene-Maurice Gattefosse, Translated from the French, Edited by Robert B. Tisserand, The C. W. Daniel Company Ltd., 1993, Reprinted 1995.

18. **Healing Oils Healing Hands**, Linda L. Smith, RN, MS, CCA, HTSM Press, 2003.

19. **Healing Oils of the Bible**, David Stewart, PhD, a CARE Publication, 2002.

20. **Integrated Aromatic Medicine**, 1998, Translated form the French and published by Essential Science Publishing.

21. **Integrated Aromatic Medicine**, 2000, Translated form the French and published by Essential Science Publishing.

22. **Integrated Aromatic Medicine**, 2001, Translated form the French and published by Essential Science Publishing.

23. **Medical Aromatherapy, Healing with Essential Oils**, Kurt Schnaubelt, PhD, Frog, Ltd., 1998.

24. **Mold and Real Estate, A Handbook for Buyers & Sellers**, Carmel Streater, Phd, DREI, South-Western, a division of Thomson Learning, Inc., 2004.

25. **Mold, Fire, Flood, & Other Topics, Homeowners Insurance Explained**, R. A. Martinez, Mission Claims, 2003.

26. **Mold Warriors**, Ritchie C. Shoemaker, MD, Gateway Press, 2005.

27. **My House Is Killing Me**, Jeffrey C. May, The Johns Hopkins University Press, 2001.

28. **Natural Home Health Care Using Essential Oils**, Daniel Penoel, MD and Rose-Marie Penoel, An American English translation from the French Original, Edited by Brian Manwaring, Distributed in North America by Essential Science Publishing, 1998.

29. **Plant Aromatics: Oral and Dermal Toxicity of Essential Oils and Absolutes**, Martin Watt, Atlantic Institute of Aromatherapy (AIA), 1995.

Bibliography

30. **Practical Art of Aromatherapy**, Deborah Nixon, Crescent Books, Distributed by Random House, 2000.

31. **Reference Guide for Essential Oils**, Compiled by Connie and Alan Higley, Eighth Edition, Abundant Health, 1998-2004.

32. **Reference Guide to Precautions in the Use of Aromatic Extracts from Plants**, Martin Watt, Balckmore, United Kingdom, 1998.

33. **Releasing Emotional Patterns with Essential Oils**, Carolyn L. Mein, D.C., Vision Ware Press, 1998-2001.

34. **Saving Face, The Scents-Able Way to Wrinkle-Free Skin**, Sabina DeVita, EdD, RNCP, The Wellness Institute of Living and Learning, Brampton, ON, Canada, 2003.

35. **Sent to Health and Annoint**, Linda L. Smith, RN, MS, CCA, HTSM Press, 2004.

36. **The Basic Chemistry of Aromatherapeutic Essential Oils**, E. J. Bowles, Sydney, Australia.

37. **The Complete Guide to Aromatherapy, The Perfect Potion**, S. Battaglia, PTY Ltd., 1995.

38. **The Illustrated Encyclopedia of Essential Oils, The Complete Guide to the Use of Oils in Aromatherapy and Herbalism**, Julia Lawless, Element Books Ltd, 1995.

39. **The Mold Survival Guide for Your Home and for Your Health**, Jeffrey C. and Connie L.May, The Johns Hopkins University Press, 2004.

40. **The Practice of Aromatherapy**, Jean Valnet, MD, edited by Robert Tisserand, Healing Arts Press, 1990.

41. **The World of Aromatherapy**, An Anthology of Aromatic History, Ideas, Concepts and Case Histories, Edited by Jeanne Rose & Susan Earle, Frog, Ltd., 1996.

42. **375 Essential Oils and Hydrosols**, Jeanne Rose, Frog Ltd., 1999.

43. **What Every Home Owner Needs to Know About Mold (And What To Do About It)**, Vicki Lankarge, McGraw-Hill, 2003.

```
┌─────────────────────────────────────────────────┐
│ ┌───────────────────────────────────────────────┐ │
│ │                                               │ │
│ │              Resources                        │ │
│ │                                               │ │
│ └───────────────────────────────────────────────┘ │
└─────────────────────────────────────────────────┘
```

Mold Remediation Supplies, Books, Diffusers, Personal Protective Equipment, Therapeutic Grade Essential Oils and oil-enhanced products, Videos, and Personalized Websites, available from:

EJC Enterprises
PO Box 368
Jackson, MO 63755
877-756-6753
Fax: 573-243-1702
info@MoldRx4u.com
www.MoldRx4u.com

Abundant Health
(Aromatherapy Books, Videos, and Supplies)
1460 North Main St., #9
Spanish Fork, UT 84660
888-718-3068
866-412-3930 (Toll Free)
801-798-0642
Fax: 877-568-1988
orders@abundant-health4u.com
www.abundant-health4u.com

AmIAQC
(IAQ Certifying Board)
American Indoor Air Quality Council
Post Office Box 11599
Glendale, Arizona 85318-1599
800-942-0832
623-582-0832 (Local)
Fax: 623-581-6270
www.iaqcouncil.org

CARE
(Essential Oils Training, Books, Videos, Audios)
Center for Aromatherapy Research & Education
RR 4, Box 646
Marble Hill, MO 63764
800-758-8629 (Toll Free)
573-238-4273
care@raindroptraining.com
www.RaindropTraining.com

ESP
(Essential Oils Books and Videos)
Essential Science Publishing
1216 South 1580 West, Suite A
Orem, UT 84058
800-336-6308 (Toll Free)
801-224-6228
Fax: 801-224-6229
info@essentialscience.net
www.essentialscience.net

IAQA (Training for Mold Inspection)
Indoor Air Quality Association
12339 Carroll Avenue
Rockville, MD 20852
301-231-8388
Fax: 301-231-8321
iaqa@aol.com
www.iaqa.org

Institute for Energy Wellness Studies
(Essential Oils Training, Books, Videos, Audios)
7700 Hurontario St., Suite 408
Brampton, ON, Canada L6Y 4M3
(905) 451-4475
info@energywellnessstudies.com
www.energywellnessstudies.com

ISHA
(Essential Oils Training, Books, Videos, Audios)
Institute of Spiritual Healing and Aromatherapy
P.O. Box 741239
Arvada, CO 80006
Fax: (303) 467-2328
Phone: (303) 467-7829
staff@ishahealing.com
www.ishahealing.com

ISIAQ
International Society of Indoor Air Quality and Climate
(An international, multidisciplinary, scientific organization. Publishes the Journal: Indoor Air)
ISIAQ Secretariat
Attn: Ms Helka Backman
PO Box 25
FIN-02131, Finland
+358 9 4355 5612
Fax: +358 9 4355 5655
info@isiaq.org
www.isiaq.org

NACHI
National Association of Certified Home Inspectors
1750 30th Street
Boulder, CO 80301
Preferred method of contact is email:
        FastReply@nachi.org
www.nachi.org

NICB
(Flooded Vehicle Database)
National Insurance Crime Bureau
1111 E Touhy Ave, Ste 400
Des Plaines, IL 60018
800-447-6282
847-544-7000
www.nicb.org

Pacific Institute of Aromatherapy
PO Box 6723
San Rafael, CA 94903
415-479-9120
415-479-0614
osa_pia@yahoo.com
www.pacificinstituteofaromatherapy.com

State of Texas Department of Health Certifications
Environmental Education Associates, Inc.
(IAQ Training. A reputable company that certifies individuals for work in the State of Texas)
346 Austin Street
Buffalo, NY 14207
info@environmentaleducation.com
www.environmentaleducation.com

YLEO
(Therapeutic Grade Essential Oils and Training)
Young Living Essential Oils
3125 West Executive Parkway
Lehi, UT 84043
800-371-3515
www.YoungLiving.com

INDEX

About the Authors

Edward R. Close, **PhD, PE,** is a recognized expert in environmental science, has served as environmental advisor to more than fifteen Fortune 500 companies, and has more than forty years experience in the environmental field. He is the author of numerous technical papers and five nonfiction books, as well as the DVD: *Toxic Mold – A Breakthrough Discovery.*

Ed holds degrees in math, physics and environmental science and engineering. He studied physics, math, philosophy, and creative writing at Central Methodist College, receiving his Bachelor of Arts degree in Math and Physics in 1963. He spent one year in the graduate physics program at the University of Missouri at Rolla and one year in the environmental engineering PhD program at Johns Hopkins University in Baltimore, Maryland. Additional studies were completed at UCLA, UC Davis, Case Western Reserve and elsewhere. He completed his thesis and received his PhD in environmental science and engineering in 1988.

Ed has more than forty years' experience in environmental planning and management, engineering, hydrology, hydrogeology, and industrial-waste management with the U.S. Geological Survey (USGS), 1965-1978, and private consulting firms, 1978-1995. While working as a research hydrologist in the Water Resources Division of the USGS, he was chosen from hundreds of employees, nationwide, to become one of the seven scientists selected to form the first Department of Interior, Systems Analysis Mathematical Modeling Group, where he worked with internationally known environmental mathematicians, including Dr. Nickolas Matalas, Dr. John Bredehoft,

and Dr. Benoit Mandelbrot.

`In 1995, Ed opened Close Environmental Consultants in Southeast Missouri and continues to serve clients that range from Fortune 500 companies, mid-size and small local businesses to individual property owners. He has worked in eleven U.S. States, on the island of Puerto Rico, and in the Kingdom of Saudi Arabia.

Ed is a member of numerous professional societies, including the Indoor Air Quality Association (IAQA), the National Society of Professional Engineers, the American Water Resources Association, the National Water Well Association, the American Institute of Hydrology, the Cape Area Engineers, and MENSA. He is a Registered Professional Engineer (PE) in the State of Missouri, a Registered Environmental Site Assessor, a Registered Well Installer, and a Registered Professional Hydrologist.

Ed has on-going interests in language, linguistics, symbolic logic, and consciousness studies. One of his books, *Transcendental Physics* (1997), explores the interface of modern physics and consciousness. He grew up on small farms in the Missouri Ozarks, where he acquired a deep respect and appreciation of nature. He is an avid outdoorsman, enjoying hiking, exploring, and horseback riding. Another of his published books, *Big Creek – History, Folklore and Trail Guide* (2003), recounts his adventures exploring caves, abandoned farm sites, and Native American villages in south-central Missouri. While working with the USGS in Puerto Rico, he served as the expedition hydrologist with a group of ten scientists on the second-ever float trip down the Rio Tanama. The expedition went underground, floating through caves five times, while observing rare species of orchids, unusual geologic formations, and giant freshwater shrimp.

Jacquelyn A. Close, **RA**, is one of fewer than 150 Registered Aromatherapists in North America. She is a member of the National Association of Holistic Aromatherapists (NAHA) and the International Aromatherapy Association (IAA). Her introduction to essential oils occurred in 1995, in a small herb shop (Cheryl's Herbs) in St. Louis, Missouri, where -- in less than three minutes --Jacqui experienced the complete cessation of her chronic and debilitating allergy symptoms. That literally opened her eyes to the power of therapeutic grade essential oils, and she immediately began an intensive study of essential oils, purchasing every book, attending every educational program and seminar she could find, and trying oils from many different companies. After completing years of hands-on training, she turned her attention to sharing with others the powerful information she had gleaned from her studies about the safe use of pure, healing, therapeutic grade essential oils.

In addition to approximately 200 hours in classes and seminars with various teachers, Jacqui has completed more than 300 hours (1999-2004) in training with D. Gary Young, ND, an internationally recognized authority on aromatherapy, essential oil production, and the originator of Raindrop Technique. She has more than 500 hours in training with David Stewart, PhD, DNM, IASP, President of CARE (the Center for Aromatherapy Research and Education). She served as a member of the CARE faculty, 2001-2006, and was a Fully Certified CARE Instructor (FCCI) from 2005-2006. In 2006, she completed nearly 100 hours in training with Linda L.

Smith, RN, MS, Certified Clinical Aromatherapist (CCA), and President of the Institute of Spiritual Healing and Aromatherapy (ISHA). In addition to her passion for learning about essential oils, Jacqui also served as President of Close Environmental Consultants, 1997-2005, and as the Director of the Conscious Healing Institute, a health and wellness center, 2001-2006. She has taught classes and seminars through the Continuing Education Department at Southeast Missouri State University and throughout North America (2001-present). Jacqui became a Registered Aromatherapist (RA) in 2001.

Jacqui and Ed began looking at complimentary and alternative health options when she had cervical cancer in 1985 and was given less than two months to live. Prayer, meditation, and the use of positive-healing affirmations brought the cancer into remission. When God brought essential oils into Jacqui's life, in 1995, she felt driven to learn as much as possible about them and credits essential oils for keeping her cancer-free. She says everything in her life has been preparation for the work God had planned for her: teaching and sharing information about the benefits of essential oils and touching people's lives with a message of healing and hope. This work has given her more joy than she could ever have imagined.

Ed and Jacqui Close have been married more than thirty years. They currently reside in Southeast Missouri, and have one son and daughter-in-law who live nearby. Since October, 2005, they have devoted a large portion of their time to research, teaching, writing, and speaking about their discovery that essential oils offer a new, non-toxic means to reduce and eliminate the risks posed to health and property by mold.

Other Books
by Edward R Close, PhD

The Book of Atma (1977)

Infinite Continuity (1987, Out of Print)

Transcendental Physics (1997)

Big Creek — History, Folklore and Trail Guide (2003)

On DVD
by Edward R Close, PhD
Toxic Mold —A Breakthrough Discovery, Saving Health and Home

This myth-busting, no-hype video presents scientific data that a safe, non-toxic blend of therapeutic grade essential oils destroys toxic mold in as little as 24-hours and keeps it from coming back. Filmed before a live audience in February, 2006, this one-hour video is the very first presentation of the data collected by Dr. Close that shows the power of essential oils in the battle against mold and the dangers mold poses to health and property.
ISBN 0-9785616-0-0

This is huge! I just watched the DVD, and its wonderful. I immediately started diffusing with the essential oil blend recommended. I really like it!
Well done. -- Vicki Opfer, Colorado

Books and Videos

by Edward R. Close and Jacquelyn A. Close
are available from quality bookstores everywhere,

and also from:

EJC Enterprises
PO Box 368
Jackson, MO 63755
Phone: 877-756-6753
email: Info@MoldRx4u.com

Quantity Discounts are available

Visit us on the Web at

www.MoldRx4u.com

Also available on the Website:

Personal Protective Equipment (PPE)
Therapeutic Grade Essential Oils and Oil-Enhanced Products
Cold-air diffusers

Now you can be assured that you are getting exactly the same essential oils, diffuser, and the protective equipment used by Dr. Edward Close in the tests he performed and reported in this book.

SAVING HEALTH AND HOME — A BREAKTHROUGH DISCOVERY

THE POWERPOINT PRESENTATION
FULL COLOR (45 Minute) PRESENTAION
and

ONE-ON-ONE

PRESENTATION BOOKLET

NOW AVAILABLE
INTERNET DOWNLOADS
at
www.MoldRx4u.com

NOW You Can Share This Information with Others
JUST THE WAY DR. AND MRS. CLOSE DO

**POWERPOINT PRESENTATION INCLUDES
THE PATIO TEST
DATA FROM THREE CASE STUDIES
THE 10-STEP ESSENTIAL OILS PROTOCOL
COMPLETED BY DR. ED CLOSE
AND SHOWN AT THE 2006 YOUNG LIVING CONVENTION**

EDUCATIONAL PROGRAMS/SEMINARS: Use this 45 minute professionally designed PowerPoint presentation on your computer, your laptop, and with a projector to explain the benefits of essential oils and their unmatched abilities to stop mold (and the health issues associated with mold exposure) and to prevent mold from coming back. Photocopy Chapter 7 of this book: "An Educational Program You Can Do," for your handouts.

ONE-ON-ONE PRESENTATION BOOKLET: When you meet with people, one-on-one, or in small groups, print the One-on-One Presentation Booklet, slip the pages into protective covers, and place them in a 3-ring binder. Then discuss this potentially life-saving information with others while sitting at your kitchen table, at a desk in an office, or at a dinner meeting. Handouts are included with the booklet.